# Glass Engineering: Design Solutions for Automotive Applications

# Other SAE books of interest

**Automotive Carbon Composites:
From Evolution to Implementation**
By Jackie D. Rehkopf
(Product Code: T-124)

**Precious Materials Handbook**
By Dr. Matthias Grehl
(Product Code: B-944)

**Design Review Based on Failure Modes (DRBFM) and
Design Review Based on Test Results (DRBTR)
Process Guidebooks**
By Bill Haughey
(Product Code: PD251136)

# Glass Engineering: Design Solutions for Automotive Applications

By Lyn R. Zbinden

Warrendale, Pennsylvania
USA

 INTERNATIONAL™

400 Commonwealth Drive
Warrendale, PA 15096

E-mail: CustomerService@sae.org
Phone: +1.877.606.7323 (inside USA and Canada)
+1.724.776.4970 (outside USA)

Fax: +1.724.776.0790



**ISBN 978-0-7680-7999-9**
**SAE Order Number R-433**
**DOI 10.4271/R-433**

**Library of Congress Cataloging-in-Publication Data**
Zbinden, Lyn R.
  Glass engineering : design solutions for automotive applications / by Lyn R. Zbinden.
        pages cm
      "SAE order number R-433"—Title page verso.
    ISBN 978-0-7680-7999-9
    1. Automobiles—Windows and windshields—Design and construction.
  2. Glass manufacture.  I. Title.
    TL256.5.Z35 2014
    629.2'66—dc23
                                                        2013048406

Information contained in this work has been obtained by SAE International from sources believed to be reliable. However, neither SAE International nor its authors guarantee the accuracy or completeness of any information published herein and neither SAE International nor its authors shall be responsible for any errors, omissions, or damages arising out of use of this information. This work is published with the understanding that SAE International and its authors are supplying information, but are not attempting to render engineering or other professional services. If such services are required, the assistance of an appropriate professional should be sought.

**To purchase bulk quantities, please contact:**
SAE Customer Service
E-mail:   CustomerService@sae.org
Phone:   +1-877-606-7323 (inside USA and Canada)
            +1-724-776-4970 (outside USA)
Fax:      +1-724-776-0790

**Visit the SAE International Bookstore**
# HTTP://BOOKS.SAE.ORG

# Dedication

To my Dad

# Table of Contents

# Foreword

In 2010, a group of body engineers began working on a concept vehicle to replace the existing bus that had been in service for about forty years. We would joke that a bus is simply a collection of glass; seats; plywood floors; and federal, state, and local regulations.

We had a team that understood plywood flooring, metal fabrication, and, of course, all the regulations. We really needed help with glass. In our search we started to utilize suppliers to aid in the creation of glass surfaces, but we didn't understand the position we were painting ourselves into. In our search for additional resources to meet the project timeline we ran into Lyn Zbinden.

Lyn had previous experience with glass from his years at General Motors, and was quick to explain how we had driven the cost of our windshield up by increasing the chord length of the glass to where it would not fit into a high-volume glass furnace. Our team had increased the glass length to avoid blind spots at the sides of the windshield. The glass suppliers were good at pointing out the need for gentle radii at the corners and a lot of other factors that made the glass look clean in its design, but had not told us that the price would increase so dramatically.

After working with Lyn to understand the issue, I asked him to document his knowledge and put together a best-practice document that summarized his years of experience. I never thought we had enough for even a hundred pages of material when he accepted the assignment. This journey has proven that you don't know what you don't know. We really had a limited knowledge of glass altogether.

Lyn has written a wonderful summary of how glass design should be approached as well as how understanding cost drivers and negotiation can make a successful product launch. His book has enabled him to simplify theory into a practical application useful to both new and experienced engineers.

Matt Thomas<br>PDT Leader Navistar 2008–2011

# Preface

My intent in writing this book is to develop engineers specializing in glass design (windshield, side, and backlite) for the transportation industry. The key word is "develop." You do not have to be a glass engineer or designer; in fact, any person with some mechanical reasoning ability can be successful.

Thirty years of experience has laid the groundwork for some practical and cost-effective engineering solutions. These design principles have been made design rules by corporations such as General Motors and Navistar. The results have been staggering: reduction in serial process design by over 75%, and warranty reduction that totals in the millions.

I will give some technical background just to refresh your physics, but the goal is to keep things practical.

My clients hire me to get their staff up and running within six months. The key to achieving success in a program remains in knowledge and the art of compromise.

Ergonomics and Aerodynamics departments (maybe managers?) support an agenda that achieves their goals. But if they are unaware of the financial constraints attached to their desired objectives, engineers may go well into the program timeline before they realize that these expectations may add more burden than what is allowed for in the program budget.

Another consideration relates to state and federal regulations which, if not thoroughly understood, may severely impact both timing and cost. This is a critical area of study if you happen to be building school buses.

An important fact that will impact or drive design will be the mode of transportation and the associated production runs. If one is working in the heavy truck, bus, or RV field, he or she will soon learn that the opportunities for favorable pricing might just have gone out the proverbial window. Limited production reduces the economies of scale dramatically, unlike the automotive industry where your production runs are based off of multiple vehicle lines and will likely be in the 150,000-vehicle range. The engineer is now faced with a dwindling supply base, unwilling to have up-front engineering work done at the supplier's expense.

I have seen many great concepts go by the wayside due to volumes and capacity constraints.

Another important factor, especially in glass, is logistics. When the purchasing department is looking to get their best price, as they should, unrealistic desires are placed upon the process. Transportation costs are the first thing that upper management keys in on during the program planning stage.

In the glass industry, float plants are few and far between. They are costly to build and must be run continuously for years to recoup the investment to construct them. In program sourcing for a product that may be assembled in the Mid-West, I may get favorable piece pricing from a supplier in the Far East. Now the fun starts.

The suppliers give you a price with a tremendous price delta, but are unwilling to set up a satellite facility near my assembly plant. Now you find that those great prices just evaporated, and you are back to the sourcing exploration phase. This is time that, unfortunately, is lost. In some large automotive companies, every minute a vehicle assembly is not operational can cost up to $26,000.

What is the purpose of the windshield in a modern vehicle? Is it a weather barrier, a projectile shield, a bug screen? In reality, a windshield is all those things and more, including a safety device. It is as important to your security while on the road as a seat belt, child seat, or even the air bag.

The windshield is designed to keep occupants from being thrown from the vehicle in case of impact, to support the roof in a rollover accident, and to position the passenger-side airbag for the most effective protection. Properly bonding that windshield to the vehicle's body is paramount to the safety of the occupants.

When I lecture or teach, I try to keep the subject light. I find if I make my students at ease with the topic, they become more engaged in the process. Unfortunately, some elements, such as the study of ceramics, can only be made so humorous.

Let's start looking at the process of making glass for ground vehicle transportation, specifically in the automotive arena of cars and trucks.

# Acknowledgments

The following companies have supplied me with guidance and cooperation:

Guardian Glass Company

Pilkington–Nippon Sheet Glass

Dow Chemical

AGC Asahi Glass

General Motors Corporation

Saturn Corporation

Navistar, Inc.

Ford Motor Company

Sekurit–Saint Gobain

Glass Association of North America–GANA

# Chapter 1
## Ceramics, Raw Materials, and Processing

## 1.1    Materials

Glass/ceramic engineering and design is the science and technology of creating objects, such as windshields and back and sidelites, from inorganic, nonmetallic materials for transportation purposes. This is done either by the action of time or temperature. Glass ceramics may have an amorphous or glassy structure, with limited or short-range atomic order. They are either formed from a molten mass that solidifies on cooling, formed and matured by the action of heat, or chemically synthesized at low temperatures using, for example, hydrothermal or sol-gel synthesis.

The special character of ceramic materials gives rise to many applications in materials engineering, electrical engineering, chemical engineering, and mechanical engineering. As ceramics are heat resistant, they can be used for many tasks for which materials such as metals and polymers are unsuited. Ceramic materials are used in a wide range of industries, including mining, aerospace, ground transportation, medicine, refinery, food and chemical industries, packaging science, electronics, industrial and transmission electricity, and guided light-wave transmission. Throughout the book, the two terms, glass-ceramics and glass, will be used interchangeably.

Glass-ceramic materials share many properties with both glass and ceramics. Glass-ceramics have an amorphous (glassy-like) phase and one or more crystalline phases, being produced by a so-called "controlled crystallization," which is typically avoided in glass manufacturing.

In the processing of glass-ceramics, molten glass is cooled down gradually before reheating and annealing. In this heat treatment, the glass partly crystallizes. In many cases, so-called "nucleation agents" are added to regulate and control the crystallization process. Because there is usually no pressing and sintering, glass-ceramics do not contain the volume fraction of porosity typically present in sintered ceramics.

The term glass-ceramic mainly refers to a mix of lithium and aluminosilicates, which yields an array of materials with useful thermomechanical properties. The most commercially important of these have the distinction of being impervious to thermal shock. The negative thermal expansion coefficient (TEC) of the crystalline ceramic phase can be balanced with the positive TEC of the glassy phase. At a certain point (~70% crystalline), the glass-ceramic has a net TEC near zero. This type of glass-ceramic exhibits excellent mechanical properties and can sustain repeated and quick temperature changes up to 1832°F (1000°C). A near-zero balance is the manufacturer's desired outcome.

Now that you have gained an insight into the basics of ceramics engineering, let's see how we can make it work in real-life applications.

The earth is rich with raw materials; let's look at what our subject consists of. Modern glass is made up of four major ingredients: sand, soda ash, dolomite, and limestone. Each of these materials contributes to the success of the manufacturing process.

Silica sand contributes as much as 70% of the glass recipe. Soda ash, the most expensive ingredient in glass, provides easier melting of the raw materials. Dolomite provides better working and weathering properties. Limestone adds to the durability of the finished product. Most glass manufacturers also use recycled glass, called cullet (15% recycled glass), in the manufacturing process. In addition to reducing material costs, cullet also melts at a lower temperature, thus reducing energy costs.

The materials just mentioned are mixed at one end of the float oven. Roughly three to four days later, a flat glass sheet of a specified thickness emerges. Once the glass is manufactured and cut to shipping sizes, it is sent to be fabricated. Generally, the float plants and the fabrication plants are in different locations. This early step of the manufacturing process is where a company looks to contain costs before the contract is awarded to a supplier. Many suppliers are eliminated from the process due to the distance between the float plant and the downstream fabricators and assemblers. The initial question from logistics, looking to reduce the overall cost in the value chain, is "Can we move the processing (float) plant closer to the fabrication plant or closer to the customer via a satellite facility?" Float plants run continuously for years just to recoup their initial investment. So, when the new engineer is engaged with his or her commodity management team and this subject comes up, processing plants are not going to move.

Let's look into the specific process in a little more detail.

## 1.2    Manufacturing Process

Two methods of manufacturing process are used in making glass: the sheet method and the float method. To make glass using the sheet method, the glass is vertically stretched, in the molten state, to the desired thickness. Because the quality capabilities are very limited with this process, today this flat glass method of production is found mostly in less-developed countries.

The float method is by far the most widely used process, today. It starts with the raw materials, mixed and melted in a refractory tank. Once the molten glass is melted and refined, it is forced through a narrow opening onto a bed of molten tin.

## 1.2.1    The Float Process

At the heart of the world's glass industry is the float process, which was invented by Sir Alastair Pilkington in 1952 in the United Kingdom. It manufactures clear, tinted, and coated glass for buildings, and clear and tinted glass for vehicles. The process, originally able to make only 6-mm (0.236-in) thick glass, now makes it as thin as 0.4 mm and as thick as 25 mm. Molten glass, at approximately 1832°F (1000°C), is poured continuously from a furnace onto a shallow bath of molten tin. It floats on the tin, spreads out, and forms a level surface. Thickness is controlled by the speed at which the solidifying glass ribbon is drawn off from the bath. After annealing (controlled cooling), the glass emerges as a "fire" polished product with virtually parallel surfaces.

## 1.2.2    Capacity

There are approximately 260 float plants in the world, with a combined output of about 800,000 tons of glass a week. Pilkington, which was purchased by Nippon Sheet Glass Co., Ltd., operates 25 plants and has an interest in another ten.

A float plant, which operates nonstop for between 11 and 15 years, makes around seven million yards of glass a year, in thicknesses of 0.4 mm to 25 mm and in widths up to 3 m. The glass is called a ribbon, much like taffy. It is continuous, with no breaks and no stoppages. The float process has been licensed to more than 40 manufacturers in 30 countries.

A typical processing flow, which follows insertion of raw materials to the loading dock, is an incredible process to watch. At one end you have raw materials being heated to ~2804°F (~1540°C); three quarters of the way down the line you watch angled cross-cutters cutting perfectly straight sheets of glass that are later divided to sizes for architectural, automotive, heavy truck, bus, windshields, side glass, and backlites.

At the onset of the material flow process, the raw materials are dumped into a hopper-like device. This then starts the journey. Glass is lighter than molten tin, so it floats, much like oil on water. The glass is slowly cooled (annealed) to a temperature of 1130°F (610°C), where it becomes rigid and is lifted off the molten tin by a series of lift-out rollers. This allows the glass to be cut and shaped later.

Note: There is an ongoing debate as of this writing on whether tin side out or air side out enables better windshield wiper performance.

Now that you have an understanding of what it takes to begin your design, let's explore the possibilities.

# Chapter 2
## How to Use What We Have Learned

The new glass engineer faces an overwhelming task of trying to balance the internal forces in the company: Ergonomics (Ergo), Aerodynamic (Aero), Purchasing, and Service.

For example, let's say the engineer has a windshield surface that meets most of the demands from the internal customer, but Ergo wants to reduce obscuration. To achieve this, the degree of view from the driver toward the passenger side of the vehicle has to be increased. A common method of achieving this is to "pull" the "wings" (A-Pillar area) back toward the cockpit. In doing so, at least two challenges are created. One, the depth of bend increases and two, distortion may be more pronounced due to trying to get the glass curvature to accelerate too fast. Ergo may be happy with this, but your Chief Engineer is not, as the piece price has doubled and tooling has increased.

Pre-formed (flat stage) dimensions should be considered before bending. Shape complexity influences price, and may affect quality, manufacturing process, and yield. Some manufacturers will try to run deep bends on the higher throughput furnaces; quality may suffer as heat is applied evenly over the entire area, and not centralized to the specific bending zone. Distortion will occur, and if this component actually makes it through plant inspection, odds are high that this will generate unmet expectations. Distortion is a key factor in high customer dissatisfaction, and will likely trigger disorientation for the driver or passenger.

Distortion is generally not much of a problem as the driver looks across his or her field of view. It is a problem, however, in areas such as the lower A-Pillar area of a vehicle. What drives this are competing departments striving to achieve their own design goals. Groups such as Ergonomics will want minimal obscuration with a bus driver trying to maintain undistorted visual contact with school children. Aerodynamics wants an accelerated curve in the pillar area to get better results in the wind tunnel, which will aid in meeting fuel requirements. Industrial Design (Styling) is trying to establish a desired shape to attract more buyers.

One of the biggest challenges faced in this area is lack of understanding of how the shape of the windshield will actually affect the total life cycle of the vehicle. Make use of internal history, lessons learned, warranty data, and more-experienced colleagues. If the design is not well thought out, and without a well-defined Design Failure Mode Effects Analysis (DFMEA) plan in place, the vehicle has a good chance of becoming problematic.

Many vehicles are produced daily without well-defined DFMEAs. Some vehicles have very aggressive shapes that may look good in the styling studio but can turn bad in real life. One vehicle I helped design had a very challenging windshield sweep. The pillar lower area was very fast (a term used in design to indicate a rapid rate of change to the surface). This would lead to the driver seeing the lines in the road and would result in a trip to the dealer. The dealer would have to replace the windshield, thus incurring a warranty charge. Keep in mind that the geometry has not changed due to the opening, the customer left somewhat satisfied, and the dealer was reimbursed by the parent company. But a warranty claim was initiated.

Warranty claims can run into the multi-millions of dollars and ruin the brand's reputation. Shape is a cruel mistress. It can garner sales as well as warranty claims. It can create challenges in packaging, particularly in drop-down glass (i.e., passenger-side windows). Its curvature can aid in strength, helping pass roof crush tests, and its flatness can produce lower tooling costs. Its flatness, as in the case of a bus, can also reduce strength but aid in reducing visual glare.

All have valid arguments and objectives to meet, but the total program has a predefined budget. This will, in the end, determine what has to be sacrificed and what will prevail.

Not only understand the desired goal and the design requirements specified to meet or exceed that goal, but fully understand your suppliers' capabilities as well as your own manufacturing plant's ability and quality history that are critical in meeting desired customer satisfaction.

There are many phases in vehicle design, and your success or failure will depend on your communication. Understand the overall program desired state. Then organize or participate in weekly or daily meetings. Get on the design review agenda, no matter how scared you may be, especially if you are new. You always need a second set of eyes. If you have an idea, sketch it out; do your homework. Challenge the accepted norm. If they say "No," say "Why?" To quote Helmuth von Moltke, "No plan of battle ever survives contact with the enemy."

# Chapter 3
## Glass and Thermal Stress

When you have decided on what you're going to design, whether it is a laminated windshield or a tempered side or backlite, you need to understand how glass reacts to thermal stress. Glass breaks in thermal shock. If you turn on a light in a cold room, such as a freezer, nothing will happen, but if you introduce an action such as spraying water (sprinkler), most likely the glass will shatter.

This can be a problem with heated mirrors. A likely root cause if you are an Original Equipment Manufacturer (OEM) experiencing failures in the field is to analyze the glass surface for grind quality. The more likely scenario, and one not first thought of, is the heating element and how quickly it has been designed to come up to temperature to defrost.

If the component has been designed to reach a sufficient defrost temperature too quickly, it will create thermal shock, much like a light bulb or cookware taken out of the oven and cooled by water from the faucet.

Thermal stress is created when one area of a glass pane gets hotter than an adjacent area. If the stress is too great, then the glass will crack. The stress level at which the glass breaks is governed by several factors. Toughened glass is very resilient and not prone to failing due to thermal stress. Laminated glass and annealed glass behave in a similar way. Thicker glasses are less tolerant. Thin glass warms up more quickly because it is thin, and not subject to differences in temperature within the glass itself that cause internal stress. With thick glass, the surface may be hot, but the center may still be cold, and the difference in temperature will cause it to shatter or crack. The higher the coefficient of expansion (COE), the more the glass expands and contracts. Conversely, the lower the COE, the stronger the glass, and the less likely it will break. Glass containing wire (defrost grid in a backlite or a heated windshield) is more vulnerable, largely due to the heating element and the amount of regulated power to the area to be heated. As noted before, the edge quality of the glass can play a part as well. Glass with damaged edges will take less stress than clean-cut glass. A good clean-cut edge is the best finish, along with fully polished edges. Ground edges may not be as good. A ground edge is a series of small defects around the glass. The effect brings all the defects to an average level and may, at best, be only more predictable than a glass with more random damage.

Heat-strengthened glass is used to resist higher mechanical loads and when there is a concern for thermal stress breakage. Heat-strengthened glass offers twice the strength of annealed glass.

Heat-treating, either heat strengthening or tempering, requires reheating the annealed glass to 1094–1499°F (590–815°C) and rapidly cooling it, so that a compression envelope develops around the glass surfaces and edges, with a balanced tension stress within the glass itself. The equilibrium of stresses causes the strength of glass to increase 2× or 4× that of the original glass. Heat-strengthened glass has a surface compression between 3500 and 7500 psi (2× strength increase). Tempered glass has a minimum surface compression of 10,000 psi (4× strength increase).

The same method used in increasing the strength in tempered glass can also be used in laminated applications.

In heat-strengthened laminated glass, two pieces of heat-strengthened glass (somewhere between annealed and tempered glass) are laminated together with polyvinyl butyral (PVB). PVB reduces the likelihood of an occupant exiting the vehicle in an unintended situation, as well as significantly reduces the level of ultraviolet (UV) rays that enter the vehicle, and diminishes the noise level inside.

## 3.1    Annealing

Annealing is a process of slowly cooling glass to relieve internal stresses after it is formed. The process may be carried out in a temperature-controlled kiln known as a lehr. Glass that has not been annealed is liable to crack or shatter when subjected to a relatively small temperature change or mechanical shock. The process of annealing glass is critical to its durability. If glass is not annealed, it will retain many of the thermal stresses caused by quenching, and its strength will significantly decrease.

## 3.2    Toughened Glass

Toughened glass is made from annealed glass using a thermal tempering process. The glass is placed onto a roller table, taking it through a furnace that heats it above its annealing point of about 1328°F (720°C). The glass is then rapidly cooled with forced air drafts, while the inner portion remains free to flow for a short time.

## 3.3    Tempered Glass

Tempered glass is a single piece of glass that is cooled quickly to provide increased impact resistance.

The process is similar to toughened glass insofar as time and temperature are increased or decreased, but the end result produces a component that will yield different testing results.

Toughened or tempered glass is a type of safety glass that is processed by controlled thermal or chemical treatments to increase its strength compared with normal glass. Tempering creates balanced internal stresses that cause the glass, when broken, to crumble into small granular chunks instead of splintering into jagged shards. The granular chunks are less likely to cause injury.

As a result of its safety and strength, tempered glass is used in a variety of demanding applications, including passenger vehicle windows, shower doors, architectural glass doors and tables, refrigerator trays, as a component of bulletproof glass, for diving masks, and various types of plates and cookware. Chemical toughening results in increased resistance to failure compared with thermal toughening, and can be applied to glass objects of complex shape. The disadvantage is that "toughened glass" is generally used to describe fully tempered glass. However, sometimes it is used to describe heat-strengthened glass, as both types undergo a thermal "toughening" process.

## 3.4    Types

There are two main types of heat-treated glass: heat-strengthened and fully tempered. Heat-strengthened glass is twice as strong as annealed glass, while fully tempered glass has typically four to six times the strength of annealed glass, and withstands heating in microwave ovens. The difference is the residual stress in the edge and glass surface. Fully tempered glass in the United States is generally rated above 65 MPa (9427 psi) in pressure-resistance, while heat-strengthened glass is between 40 and 55 MPa (5801 and 7977 psi, respectively).

## 3.5    Strength

A rule of thumb for glass is that its strength is a function of the square of its thickness. As an example: the square of 5 mm is 25, while the square of 4 mm is 16. So, a 5-mm part is about 50% stronger than a 4-mm part. Impact strength is not as direct a correlation because a lot has to do with mass and shape of the "projectile," but the square rule gives an idea of the thickness benefit.

It is important to note that the tempering process does not change the stiffness of the glass. Annealed glass undergoes a similar deflection compared to tempered glass under the same load. But tempered glass can take a higher load and, therefore, deflects further before breaking.

## 3.6    Processing

The disadvantage is that toughened glass must be cut to size or pressed to shape before toughening, and cannot be reworked once toughened. Polishing the edges or drilling holes in the glass is carried out before the toughening process starts. Because of the balanced stresses in the glass, damage to the glass will eventually result in the glass shattering into thumbnail-sized pieces. The glass is most susceptible to breakage due to damage to the edge of the glass where the tensile stress is the greatest, but shattering can also occur in the event of a hard impact in the middle of the glass pane, or if the impact is concentrated (for example, striking the glass with a point). Using toughened glass can pose a security risk in some situations. It does have the tendency to shatter completely upon hard impact, rather than leaving shards in the window frame.

The surface of tempered glass does exhibit waves caused by contact with flattening rollers, if it has been formed using this process. This waviness is a significant problem in manufacturing of thin-film solar cells.

# Chapter 4
## Forming for Specific Vehicle Positions

## 4.1 Laminated Glass

Laminated glass is a type of safety glass that holds together when shattered. In the event of breaking, it is held in place by an interlayer, typically of polyvinyl butyral (PVB), between its two or more layers of glass. The interlayer keeps the layers of glass bonded even when broken, and its high strength prevents the glass from breaking into large, sharp pieces. This produces a characteristic "spider web" cracking pattern when the impact is not enough to completely pierce the glass.

## 4.2 Application

Laminated glass is normally used when there is a possibility of human impact that could lead to occupant ejection. Automobile windshields use laminated glass. The PVB interlayer also gives the glass a much higher sound isolation rating, due to the damping effect, and also blocks 99% of incoming UV radiation. This has been used in high-end automobiles to reduce interior cabin noise.

Note: Laminated sidelites have only been made available to lower-cost vehicles in recent years.

## 4.3 Forming

Laminated windscreens for truck, bus/truck, and automotive applications are formed from two sheets of glass that are separated by a polymeric interlayer. They are usually curved in one or, often, two directions, making their manufacture complex.

## 4.4   Processing

The process is further complicated by the need to apply a coating to one or more surfaces of the laminate.

Several methods are used for the production of laminated windscreens. One method involves applying a coating to an inner surface of one flat glass sheet and firing the sheet to affix the coating. A second flat glass sheet is placed on top of the coated glass such that the coating is between the sheets, and both sheets are then heated to a point where shaping can be performed.

A second method involves applying a coating to an outer surface of a first flat glass sheet, and placing this glass sheet on a second flat glass sheet such that the coating remains on an outer surface. The sheets are then heated and shaped together, the heating cycle also affixing the coating to the first sheet. Once cooled, the sheets are separated and their order reversed so that the outer coated surface becomes an inner surface. Insertion and bonding of a polymer interlayer completes the process.

Again, with this method there is no adhesion or transfer of the coating, offering the advantage that fewer heating steps are required. A disadvantage is that small differences in curvature between the first and second glass sheets can cause stresses and optical distortions when the sheets are inverted to form the final laminate.

A third method involves applying a coating to the inner surface of a first flat glass sheet and drying the coating using a low-temperature heat cycle. A second flat glass sheet is placed on top of the first sheet, enabling the dried coating to remain intact between the two sheets. Both sheets are then heated and shaped, the heating cycle also affixing the coating to the first sheet. After cooling, the sheets are separated, and a polymer interlayer is inserted and bonded to form the laminate.

This process has the advantage that the low-temperature drying cycle does not lead to optical distortions in the first sheet. It is also less expensive. The method does, however, place limitations on the types of coating material that can be used successfully. As the coating is only dried and not bonded to the first glass sheet prior to shaping, adhesion and transfer may occur.

Furthermore, coating compositions contain organic vehicles to facilitate application by printing; such organic components are not completely driven off during drying. In the case of the first two methods, the organic content of the coating does not give rise to problems, as it is easily oxidized during one of the high-temperature heat cycles. This is because, at least once, the coating is on a surface exposed to the atmosphere (air) when at high temperature. In the third method, organic components are trapped between the glass sheets when at high temperature, and thus have no access to atmospheric oxygen. Nevertheless, this third method is generally preferred industrially.

## 4.5    Added Value

Additional conductive coatings may be provided for use as antennas or as resistance heaters. The coatings are most often applied to an inner surface of one of the glass sheets, making up the laminate (i.e., a surface in direct contact with the interlayer). Coating materials are usually ceramic inks or enamels, and as such must be fired at elevated temperatures to bond them to the glass.

It is important that during manufacture, the coating does not stick or transfer from one glass sheet to another. Post cooling, the sheets are separated and a polymer interlayer is inserted between the glass sheets. A final heating step bonds the laminate together. The method avoids adhesion or transfer of the coating to the second glass sheet. However, firing to affix the coating is both expensive and can lead to optical distortions in the final laminate.

## 4.6    Construction

For automotive use, a windshield is generally symmetrical in cross-sectional thickness, while the heavy truck industry needs more robustness for severe service use and specifies what is called an asymmetrical design. An asymmetrical approach to design keeps the weight the same as a normal symmetrical design but increases the resistance to cracking. Sorry, stone chips are still a challenge.

As with many components designed for the transportation industry, in automotive (including truck, bus, and passenger vehicles) there are as many differences as similarities.

## 4.7    Non-laminated Glass

Non-laminated or tempered glass is used when strength, thermal resistance, and safety are important considerations. The most commonly encountered tempered glass is that used for side and rear windows in automobiles.

It is used for its characteristic of shattering into small cubes rather than large shards, covered in ANSI Z26.1, and sometimes referred to as safety glass. This is so that, upon impact, occupant harm is lessened. (The windscreen or windshield is instead made of laminated glass, which will not shatter when broken.) Prior to using tempered glass, plate glass was used, but failure (glass breakage) would result in large shards that could cause great damage to the occupants.

Laminated glass, the type used in windshields today, has a buffer of PVB to hold the glass together.

# Chapter 5
## Design Guidelines and Applications

## 5.1  Design Guidelines

The National Highway Traffic Safety Administration Federal Motor Vehicle Safety Standards (FMVSS 205) and Economic Commission for Europe (ECE R43), along with most companies and materials engineering groups, mandate glass safety criteria. In addition to the aforementioned requirements, bus glass must meet glazing specifications dependent on applicable state regulations. Current construction for the windshield is a 3.5-mm outer lite and a 2.3-mm inner lite.

If a company were to re-spec the outer glass thickness to 2.3 mm, an estimated $\Delta$ savings would be approximately \$1.00/part, and the overall thickness would be reduced by 1.2 mm. The new windshield thickness would be similar to automotive windshields, which would still meet all the in-house material specifications, FMVSS, and ECE specs. Most sidelites should be 4.0 mm thick. It is not recommended that this part be thinned down to less than 3.1 mm. The price $\Delta$ for this type of thickness change would be about \$0.15/part. Tooling changes for this part would include a gage and injection-molded spacers.

### 5.1.1  Sidelites for Bus Applications

For school buses, reduction in stationary glass thickness (< 1/8 in or 3.048 mm) would prohibit sales in the following states: Arkansas, Florida, Kentucky, and South Carolina, where the glass thickness required is 5.58 mm. Reducing glass thickness would expose your company to a greater risk of in-transit related incidents. Replacement cost in the field is roughly \$600.00.

## 5.1.2   Key Factors

The following critical objectives must be met and will affect the success or failure of your design.

- **Tooling**: (Encapsulation, gage, contour, and bending fixtures) affected by a reduction thickness would need to be adjusted to accommodate the thinner glass.

- **Windshields**: The difference in expected breakage rate from impacts should be tested and understood.

- **Door Glass**: Testing would be needed to see if reducing the glass thickness impacts the lift mechanism, or negatively affects vibration noise. Sealing should be tested.

- **Design Failure Modes and Effects Analysis (DFMEA)**: Would need to be performed, and would likely point toward retesting much of the Design, Validation, Plan and Report (DVP&R).

## 5.1.3   Critical Parameters for a Good Design

- Certain parameters will make for a good design. These critical elements will not only dictate good performance but also better enable you to achieve your cost targets.

- Rate of change: The rate of change in horizontal or vertical curvature will affect optics and wiper performance.

- Installation angle: The installation angle will impact veiling glare.

- Depth of bend: The depth of bend may dictate your supplier choice. One glass company currently makes a windshield with an A-pillar radius of 222 mm and an angle of 75°. It would be hard to produce the windshields with angles closer to 90°. The issue you will run into is distortion if you try and get a full wrap-around to the corner.

- Distortion: A degree of distortion, when looking through, is inevitable in curved glass, particularly when viewing a moving object through the glass. It should also be noted that curved glass will split direct sunlight into striped shadow. This phenomenon will occur if you try to design your glass to do something it is incapable of doing in the area given to work with.

## 5.1.4   Curvilinear Distortion

Curvilinear distortion is the optical distortion produced as the glass deviates from a straight line. The phenomenon becomes more pronounced if the rate of change (accelerated curvature, fast curve) increases rapidly over a short distance.

There are three things to consider regarding the geometry between the viewer, the glass surface, and the object:

1. Distortion of an object far away is often greater than of an object that is near. Also, if the viewer is right up against the glass, distortion may not be evident, whereas it can be obvious if the viewer is at a great distance from the glass.

2. Whether the object is viewed through the center or periphery of the glass. As in optics, distortion often increases toward the periphery of the glass.

3. Angle of view of the object. Distortion often is more evident when an object is viewed obliquely through glass rather than at or near a 90° angle to the glass surface.

When planning a vehicle, "bucks" should be created to simulate the driving experience. A buck can be full running or static.

As shown in Figure 5.1, the driver will always have a distorted view looking outward. At best, this would be unpleasant; at worst, it would give an unrealistic view. As the driver rotates his or her head, the field of view changes, and the perception of glass thickness increases. As you can see in the figure, at the Edge of Glass (EoG) around the 27.37–mm dimension, the viewer is looking right into the glass thickness and not through a targeted line of sight. This effect can manifest in the windshield, backlite, or quarter glass area. It is caused by asking the glass to change too fast, usually trying to tie into the sheet metal at a pillar. A buck will illustrate this effect to engineering and styling before an actual prototype is built, costing many hundreds of thousands of dollars, both in the tooling, production of parts, and labor.

Many glass companies have optical simulation tools that can detect these potential effects prior to the final design phase. Unfortunately, many of them are static in nature and do not follow the dynamic movements of the viewer.

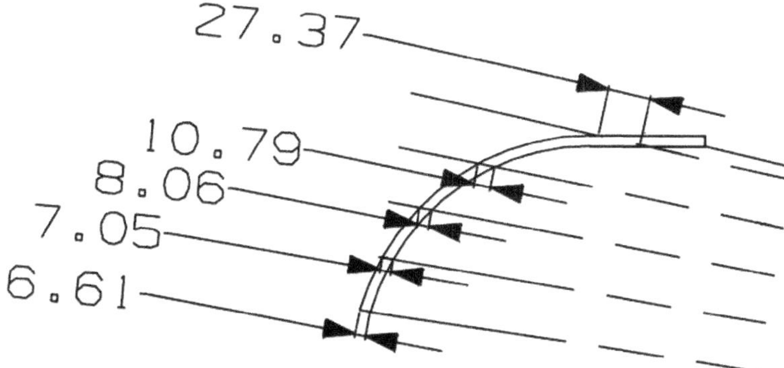


**Figure 5.1** Detail of curved glass.

## 5.1.5  Meeting SAE J198

The relationship between height and width can affect whether the manufacturer will be successful in meeting the guidelines set forth in the SAE International standard, SAE J198. This is critical for many reasons.

Most of all, if you are selling a school bus, for example, to a particular state or government agency, they may mandate that this regulation be met. You may (or not), if your design is not within tolerance, be able to sell a vehicle. In one case, this affected a manufacturer's sales, decreasing them by 33%.

## 5.1.6  Computer-Aided Engineering

Surface flow lines will show "hot and cold" ranges using Computer-Aided Engineering (CAE) simulation. The red areas are the flattest part of the windshield. As mentioned before, an accelerated rate of change will contribute to poor optical qualities as well as potential wiper chatter. Wiper chatter is caused by the wiper blade being unable to transition from out-wipe to in-wipe as the wiper blade flips when it changes direction. In some ways, wiper technology is still a dark science. Many wiper manufacturers do not specify load in Newtons (N). If it works, then it's fine. With that in mind, reduce the rate of change as much as possible when designing your surface.

You will also notice the asymmetrical lower edge of glass. This is dictated by packaging the right-hand side of the wiper assembly.

## 5.1.7  Polycarbonate Glazing

Automobile glazing is one of the fastest growing market segments of polycarbonate resins. This application is the most common in Europe, where molded polycarbonate is allowed in nearly all auto glazing functions. Polycarbonate glazing in the United States is used in side, rear, and roof panels, but not for windshields, where its use has not been approved.

Polycarbonate glazing has a number of advantages compared with glass:

- It is anywhere from 40–60% lighter in weight than glass, resulting in improved fuel economy.
- It can be injection molded into complex shapes and designs.
- It can be produced with distinctive colors and textures.
- It permits parts integration.
- It is very tough, protecting vehicle occupants in accidents.
- It offers improved security against theft because it does not shatter as easily as glass.

Although polycarbonate has been used for many years in auto headlamps, its progress in window glazing has been slowed by a number of limitations. These include the tendency of polycarbonate to scratch easily, plus its UV sensitivity, which makes it more weather vulnerable than glass. Until recently, the equipment was not readily available to produce high optical clarity in relatively large-size (up to 1.4 m$^2$) polycarbonate panels, particularly under the required low-stress injection molding conditions. Developers are enlisting various technologies to address these limitations. Polycarbonates may offer a price penalty up to 3× higher than glass. It is limited by federal safety standards in its use or position in a vehicle. To improve abrasion resistance of polycarbonate auto glazing, major resin companies have developed various coating technologies. Heat-cured wet formulations are the traditional coatings applied to small-area polycarbonate glazing. But wet coating methods don't always provide adequate scratch resistance in large-area glazing such as windshields. One newer dry coating method that offers sufficient hardness is plasma-enhanced chemical vapor deposition (PECVD) of various organic silicon coatings. In PECVD, an electric field enhances formation of ionized species in the coating chamber, allowing much lower coating deposition temperatures than other chemical vapor deposition methods. As a result, PECVD allows high rates of deposition without thermal damage to the polycarbonate.

Polycarbonate glazing must also be protected against UV light, which can turn it yellow. Hard coats of silicon are the most common approach today for UV protection.

Today's polycarbonate auto glazing systems also contain layers with adhesion promoters, and decorative and printed layers. The underlying polycarbonate can contain various dyes and additives to improve the esthetics and performance of the component.

## 5.2    Specific Design Applications

Up to now, our focus has been on designing windshields. Sidelites or side windows require the same discipline, many of the same manufacturing processes, as well as sharing, in some cases with the fixed-vent window, the same bonding characteristics. As far as styling goes, the bus window, definitely, and truck window, to some degree, is pretty straight forward and visually unappealing.

### 5.2.1    Truck and Bus Design

Truck windows will have curvature but, unlike passenger cars, their shape is not multi axis, meaning they generally will have curvature in the YZ axis. This effect, noticeable in passenger cars, produces a phenomenon called "tumblehome." If you were to look at a mail truck straight on, or normal, you would see a seemingly straight line from the ground plane up. Curvature does a couple of things for the vehicle. It gives it "style" and can improve aero-dynamics, which will result in better fuel economy, the latter being critical to volume fleet buyers who gauge good mileage in tenths of a gallon. When you, as a retail sales entity, are looking at a vehicle commanding a $160,000.00 price tag, this is of concern.

Since the glass in a side window generally is tempered and not annealed, as a windshield this also changes the manufacturing, testing, and meeting of government regulations.

### 5.2.2    Passenger Vehicles

The most critical part of designing a passenger car sidelite window system is the overall packaging envelope.

### 5.2.3    Packaging

Clearances and water tightness are critical. In a truck or passenger car, many components come into play for packaging, each one interdependent on the other.

### 5.2.4    Mass

Mass in a window system affects opening and closing efforts, packaging the motor, and, to some extent, vehicle handling because it adds to its overall weight. One must first consider mass and center of gravity of the window system itself.

If a motorized system is required, then consideration must be given to the environment. The design engineer must consider space and vapor barrier to protect a motorized mechanism, not to mention the load on the electrical system.

## 5.2.5  Careful Design

In general terms, the part will be easier to fabricate if the geometry is less complex. If it is a drop-down glass, then additional cost must be factored in for edge finish. The glass as a component is the easiest part to design. The attaching of hardware is where the opportunity to meet mass and cost goals lies.

## 5.2.6  Testing

Design initiatives such as this require a thorough knowledge of the system. Component test for the glass itself will be conducted by the supplier; system testing is performed by the Original Equipment Manufacturer (OEM).

## 5.2.7  Other Considerations

The designer must try to guide styling and aerodynamics toward a theme that not only provides a level of acceptance, but also meets program cost targets. The following sections cover more considerations when designing tempered glass.

### Shaping and Forming Sidelites

There is not a lot going on with respect to design in this case, due to packaging limitations in the door panel, etc. Sidelites that are moveable (door glass) and that have an exposed edge must meet certain standards with no variation. Variation in this case could lead to unmet design expectations with respect to physical stress, as well as customer dissatisfaction relating to the touch and feel of the product.

### Edge Stress

A common reason glass breaks is improper seaming or grinding. When you cut the glass you actually score it, and then break it out. This leaves tiny stress cracks along the breakout line.

### Finishing

The seaming and grinding removes stress cracks. The ground edge impacts the edge stress only for tempered parts. Laminated parts are not affected.

The reason for grinding is also to remove sharp edges, and the seamed edge does the same job as the ground. Additionally, the ground edge, compared to a seamed edge, will give a smoother finish and consequently will provide a much better appearance. The ground edge is used mainly on exposed edges where the glass is bonded and there are no gaskets to cover.

When designing sidelites or fixed vents, careful consideration must be paid to distances from sheet metal that may induce chipping or outright breakage. Five to six millimeters of gap is recommended.

# Chapter 6
## Processing

## 6.1    Windshield/Windscreen Applications

Newer developments have increased the thermoplastic family for the lamination of glass. Besides PVB, important thermoplastic glass lamination materials today are EVA (ethylene vinyl acetate) and TPU (thermoplastic polyurethane). The adhesion of PVB/TPU and EVA is not only high to glass but also to Polyester (PET) Interlayer. Since 2004, metalized and electro-conductive PET-Interlayers are used for light-emitting diodes (LEDs) and laminated to or between glasses.

- **Top Layer:** Glass
- **Interlayer:** Transparent thermoplastic material such as TPU, PVB, or EVA
- **Interlayer:** LEDs on transparent conductive polymer
- **Interlayer:** Transparent thermoplastic material such as TPU, PVB, or EVA
- **Bottom layer:** Glass

There are several laminated glass manufacturing processes:

1. Using two or more pieces of glass, bonded between one or more pieces of plasticized polyvinyl butyric resin using heat and pressure.

2. Using two or more pieces of glass and polycarbonate, bonded together with aliphatic urethane interlayers under heat and pressure.

3. Interlaid with a cured resin.

Each manufacturing process may include glass lites of equal or unequal thickness. Glass shaping and volume will dictate process.

Glass with a camber over 250 mm will be formed on a "top-hat" furnace. This is sometimes referred to as a single-box furnace, as the glass is processed one at a time to ensure the proper shaping and bends in the high curved area. Heat can be applied selectively to avoid distortion for higher quality. Cycle times on a top-hat furnace run up to 30 minutes per part.

Glass with a camber under 250 mm can be run on a box furnace. This is a continuous process furnace with a box moving through the heating zone. Cycle time on a box furnace runs from 1.5 to 2 minutes.

Glass with a camber under 180 mm and annual volume of 40 thousand units will be reviewed for a press bend furnace. Overall size of the glass is checked to ensure that it fits into the platen on the press. This is a very high-volume furnace with a continuous process. Cycle time on a press bend furnace runs under 1 minute.

Depth of bend (camber) and arc length: these dimensions are critical. They can define your supplier choice, yield, and cost. Not all glass companies have the same dimensional capacity. The difference in oven height can be as much as 5 in (125 mm). This can affect yield on a normal production run. This is an important factor when designing, engineering, and sourcing glass. If these dimensions are not considered from the beginning, a design may be created and go through its review process only to find out six months down the road that company A cannot, in fact, produce it for the target price. This is a very real possibility that will generally happen only with buses, trucks, and recreational vehicles. The decision to move to a larger, slower oven may impact piece price delta by > 60%.

Horizontal depth of chord, or Hdoc, is also referred to as horizontal depth of bend. This dimension is taken after the bending process, usually at the widest point of the windshield. Figure 6.1 shows measurement of Hdoc.

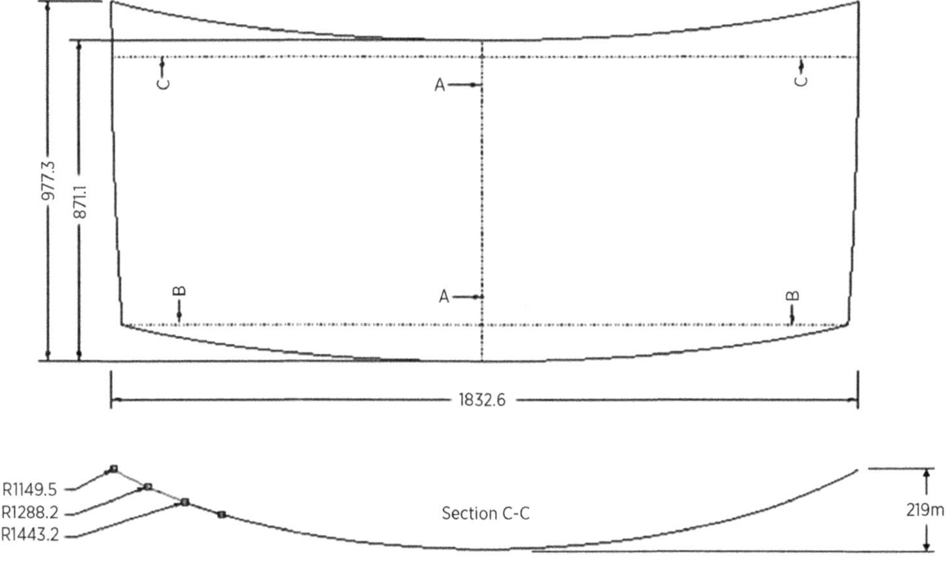


**Figure 6.1** Horizontal depth of chord (Hdoc) section.

## 6.2 Windshield and Sidelite Processing

The four types of bending processes to consider when forming laminated glass are:

- Press bending: Ultra-high flow rate and highest yields
- Lehr: High flow rate and high yields
- Box (or Lamino): Medium flow rate and good yields
- Tanglass: Ultra-low flow rate and low yield

### 6.2.1 Press Bending

This process is used for high-volume parts due to expensive tooling and long change over (minimum 15,000 vehicles/year). This type of bending accommodates more-complex bends and curvature in two planes. It enables the group responsible for industrial design and aerodynamics to achieve the desired program initiatives. Tooling piece price is more expensive than sag bending piece pricing.

This type of forming is used in all vehicle positions. It has size limitations: max. 1800 mm × 1200 mm, and 170 mm depth of bend for a windshield. Figure 6.2 shows constraints for the use of high-yield ovens.

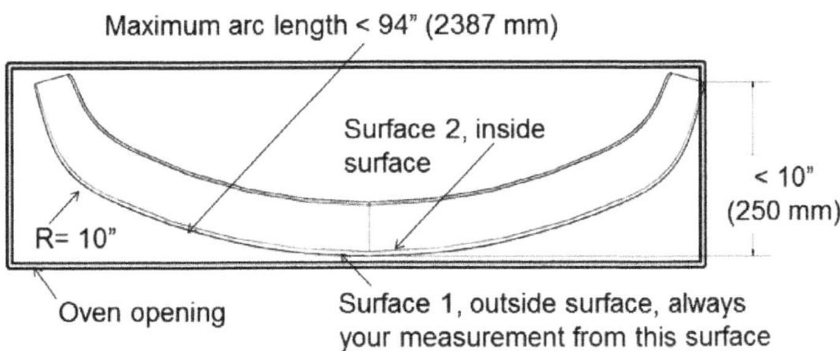


**Figure 6.2** High-yield oven constraints.

Figure 6.3 presents a windshield forming capability matrix.

**Figure 6.3** Windshield forming capability matrix.

### 6.2.2  Box (or Lamino)

This method is used for medium volumes (minimum 5000 vehicles/year). You probably will not hear of this process too often. This furnace is mostly employed by the truck market that has a low-yield application: max. 2000 mm × 1250 mm, and 250 mm depth of bend.

### 6.2.3  Tanglass

This method is used mainly for trucks and off-road vehicles, capable of producing huge parts due to flow rate limitations. It is used for a maximum production run of 5000/year. Maximum size is 3200 mm × 2100 mm, and 350 mm depth of bend.

### 6.2.4  Press Bending for Windshields or Other Laminated Position

In a press bend furnace, the depth of bend is limited to the press bend opening, height of the cooling section, and mainly the vacuum step, where the windshield is put in bags using vacuum to extract the air between the glass and PVB. The bags have a curvature limit, and if it goes over the limit, the windshield will break inside.

### 6.2.5  Sag Bending

During the sag bend process, at the furnace stage, the sag bending tool goes inside a box (thus the name of the box furnace). The glass will remain in the box until it is totally bent and cooled down; the height of the box and glass lite will be much higher than the press bend.

### 6.2.6  The Vacuum Stage

Vacuum rings are installed all around the glass to extract the air trapped between the glass and PVB. The depth of bend will not interfere in this stage.

The shape of the glass will interfere directly with the capability of the furnace, meaning that the deeper the camber, the more complex the glass, and the shorter the maximum size. Less-complex shapes lower costs and increase flow rate, resulting in higher yield.

### 6.2.7  Hardening

Tempered glass, unlike annealed glass, is quick cooled rather than slow cooled. This causes the outer surface to harden faster than the core. As the core hardens, it causes pressure on the outside surface, giving the glass additional strength to withstand the closing of the door.

### 6.2.8  Understanding Costs

The expected tooling price per process will be as follows:

- Average sag bend runs around $3000, but you use multiple tools, depending on volume.
- If the volume is 10,000 per year, you may use 15 to 20 tools.

- Average press bend (there are many types that can vary widely depending on bend and complexity) for an average complex part may run $180,000.

- Polyurethane material per part can vary from a minimum of 40 g to a maximum of 1500 g.

- This is dependent on part size, design, and geometry.

### 6.2.9  Basic Considerations

When you have chosen your type of processing, sag form or press, for a windshield or backlite, you must also consider the final chapter of the manufacturing phase, that being finishing off the edge of the glass.

Many factors are involved here, such as where the part is used, windshield, sidelite, backlite, as well molding cover; or if the part is heated.

Glass for use in motor vehicles is tempered or laminated. And its greatest point of fracture is within the first 20 mm from its edge. When a hole is drilled through a piece of glass, it is drilled prior to the strengthening process. Holes are sometimes necessary to attach the carrier mechanism for body side door. If this is the case, then tolerance stack-ups are critical to ensure that the window movement will be tightly controlled. Figure 6.4 shows stress cracks caused by the cutting process.

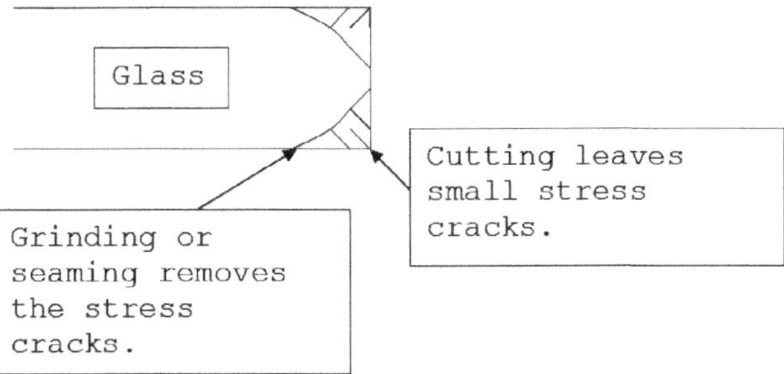


**Figure 6.4** Edge stress.

### 6.2.10  Interfaces

In many cases, the window itself will sit in an aluminum extruded channel that attaches to the window lift mechanism. This entire assembly is subject to up and down load, moisture, and vibration. Carefully consider these factors.

### 6.2.11  Load

Center of gravity will also play a key role, whether or not the system meets its desired objective. Center of gravity will play a major part in window opening and closing efforts, as well as life of the window motor. Center of gravity in a side window is very often overlooked as a potential cause for opening and closing efforts.

It is an easy thing to check using Computer-Aided Engineering (CAE). It requires no external labor, is low cost, easy to check, and easy to correct. This is the mindset a successful design engineer needs to adopt. "What can I do that is in-house, quick, and easy, and stay within my timeline and budget?" Think before you spend!

# Chapter 7
## Ceramic Frit and Shadeband

Frit is a key feature to aid in the preservation of the various components such as urethane. Frit is also called ceramic frit or a paint band. It is usually black, but it can be made in different colors and in silver, also, for the heat paint.

## 7.1 The Process

Ceramic frit is painted using a silk screen printing before it goes to the furnace. For a double printing, the first layer has to be dried. The application of ceramic frit is often helpful for protection in a variety of ways, which will be explained.

## 7.2 Composition

Ceramic frit is a paint composed of minute glass particles and pigment. The enamel is fired onto the glass using a silkscreen process, creating a permanent coating. A windshield needs frit for UV protection.

## 7.3 Artwork

Frit band can be applied with a gradient dot pattern. The smallest dot should be 0.8 mm or larger.

The Original Equipment Manufacturer (OEM) will design the artwork (develop a blackout line, dot matrix pattern). This is done not only to reduce UV degradation but also to give a more aesthetically pleasing view into the interior of the vehicle. A positive image of the artwork is created and sent back to the processing plant.

## 7.4 Supplier Capabilities

All glass manufacturers have their own screen rooms where they take the positive image, put it in a chamber, and shoot it onto a silk screen. Coatings may be decorative or functional. Often, an edge coating is used to obscure other components or to protect the adhesive used to fix the windshield into the body of the vehicle. Absent such an edge coating, the adhesive may be subject to degradation by sunlight.

The chamber shoots the positive onto the screen. Then an emulsion is laid down on the areas where you do not want the ink to go through the screen. (The emulsion blocks off the holes in the screen.)

The screen is loaded into an aluminum frame, which is then loaded in the screen printer, in line with the furnace. The image is screened on when the glass is in the flat state before the furnace. The frit then fires into the glass when it goes through the furnace.

## 7.5    Artwork Costs

The cost for tooling (shooting the artwork, making the positive, silk screens, and frames), usually applied to surface number 4, ranges between $2000 and $4000 (U.S.). The added cost per piece depends on the complexity.

A price of $3.00/part should be more than adequate. All prices listed in this book are subject to inflation, most of all concerning raw material. Piece price and artwork will increase, but not by too much. Tooling price, however, can increase dramatically. Use the prices as a go-by for model year 2014. Silkscreen changes (e.g., logos, markings, etc.) will cost approximately $1500.

## 7.6    Manufacture of Frit

Johnson Matthey, an English company, produces most of the frit used in the industry.

## 7.7    Artwork Changes

Part price is dependent on the artwork change. For adding or removing a complex shape such as a dot matrix or strobe pattern, you can estimate $1.00 increase/decrease. Large increases or decreases in paint will affect yields, and can affect price by one to two dollars. Simple, small millimeter changes here or there will be price neutral.

## 7.8    Shadebands

Stretched shade bands are only used on parts with a large smile shape at the top of the windshield. Glass manufacturers stretch the shade band to match the smile. (If the viewer were looking at the windshield in a front view, the glass edge shape would appear as a smile.) This would maintain a consistent shade band height along the top edge.

# Chapter 8
## Adhesive Bonding

To meet various tests such as roof crush, as well as occupant retention, an additional requirement put on vehicle manufacturers is to protect occupants in a crash, which means that the glass must be bonded to the body of a vehicle by a fail-safe method. This requirement was met by the introduction of polyurethane as the bonding adhesive. Urethane, as it is commonly known, is a very special moisture-attracting adhesive with tremendous bonding strength characteristics.

## 8.1 Types

In comparison, the prior glass adhesive, butyl, has approximately 30 psi of tensile strength, while urethane has approximately 1000 psi of tensile strength. This strength is necessary for windshield retention, roof support, and airbag deployment in the event of a crash.

Currently, every passenger vehicle and truck under 10,000 lb gross vehicle weight uses urethane to bond the windshields to the car body. One OEM's glass bonding specification requires minimum shear strength of 3.5 MPa (500 psi) and 100% cohesive failure of the urethane adhesive after initial exposure, with 75–95% retention of that strength after various environmental exposures. Urethane requires two things to cure: moisture and temperature.

## 8.2 Temperature

The higher the temperature and the more humidity available, the faster it will cure. Good processing requires no more than about 5 minutes open (the time after the urethane has been applied to the windshield to the time it is placed in the cab opening) time before decking (placing in cab opening), particularly in the warm summer months, or the material will skin over and will not bond once decked. The approximate maximum open time would be 8 minutes during the winter months; this is the slowest time of the year for urethane curing due to colder temperatures and lower absolute, not relative, humidities. Testing results have shown that the urethane will achieve strength of 500 MPa at 73.4°F (23°C) in three days. During the winter months it could take a week for the material to fully cure.

## 8.3 Interface Control

For proper bonding, I recommend a minimum of 6 mm gap between glass and bonding, and a minimum of 15 mm bonding path width. Urethane is dispensed onto the primed main body with the nozzle tip perpendicular to ensure uniform contact.

Press the window into the frame until the triangle bead is squeezed to half its original height. The cross-section of the bond should be approximately ¼ in × ¼ in (6 mm × 6 mm) square after application. Figure 8.1 shows minimum urethane bead height from sheet metal flange to glass surface.

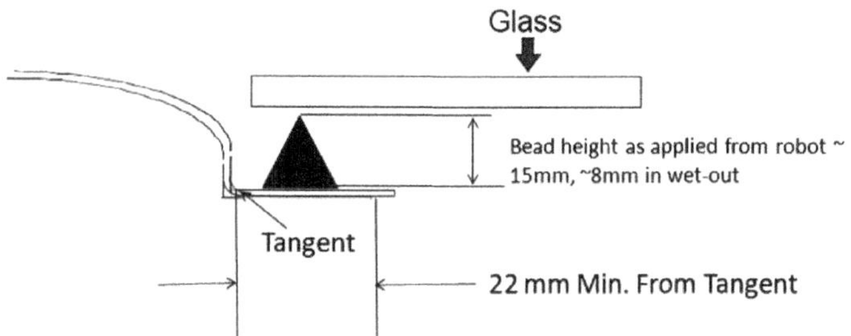


**Figure 8.1** Minimum urethane bead height.

## 8.4 Locating Scheme

A common method of locating and mounting a windshield in correct body position is an adhesive-backed pin, usually made from a plastic. The internal specification for glass bonding requires minimum shear strength of 3.5 MPa (500 psi) and 100% cohesive failure of the urethane adhesive after initial exposure with 75–95% retention of that strength after various environmental exposures.

OEM glass bonding systems are also tested on crash vehicles per FMVSS 212 requirements (for vehicles less than 10,000 lb). The main purpose is to hold the glass in place until the urethane sets up.

## 8.5 Surface Preparation

All surfaces must be clean, dry, dust free, and frost free prior to sealant application. This means removal of all dirt, dust, oils, and all other forms of surface contaminants prior to the application of any sealant materials. The following cleaning solvents may be used to remove surface contaminants:

- 50% solution of isopropyl alcohol (IPA) and water, 70% solution of IPA and water (rubbing alcohol), or pure IPA for non-oily dirt and dust.

- Methylethylketone (MEK), Xylene, or Toluene for oil and grease. IPA and MEK are soluble in water and may be more appropriate for winter cleaning as they help in removing condensation and frost.

- Use primer when required.

- Certain substrates that chalk or oxidize require primer (see manufacturer's spec sheet).

- For joints subject to water submersion, surface condition at time of application.

- Sealing to wet or frosty surfaces will cause sealant failure.

## 8.6    Urethane Adhesive Application

Adhesives are applied in bead form with a pressurized flow gun or other suitable application system. They can be applied manually or robotically.

## 8.7    Storage

Adhesives are a moisture-sensitive material and must be stored in airtight, dry containers at temperatures between 64.4°F and 95°F (18°C and 35°C). The adhesives should not be stored outside or in direct sunlight.

The shelf life of this material is assured for six months (from the date of manufacture) at temperatures not to exceed 95°F (35°C). The material is capable of withstanding a maximum temperature of 109.4°F (43°C) for up to three days. As for low temperature stability, the adhesive is capable of withstanding at least three cycles of 17 hours at –20.2°F (–29°C), followed by 7 hours at 77°F (25°C). Urethane adhesive and primer packages are marked with expiration dates.

## 8.8    Clean-up

If clean-up is required, the uncured material shall be readily removable from production topcoats, glass surfaces, glass inner liners, and vinyl top materials with vamp naphtha, or other solvent agreed upon at time of approval with appropriate engineering groups.

## 8.9    Primers

Some primers are solvent based. A Silane-blend primer is used to condition the surface and promote adhesion to glass. This primer is used in conjunction with an approved blackout glass primer and polyurethane adhesive. Figure 8.2 shows a 3-D cross-section of the layer build up from sheet metal flange to glass surface.

When applying primers, many manufacturers recommend that 3–5 mils wet (approximately 0.3–1 mil dry) be applied to the glass (1 mil = 25.4 microns). Coverage should be sufficient to obscure the presence of glass (continuous film with no skips or streaks).

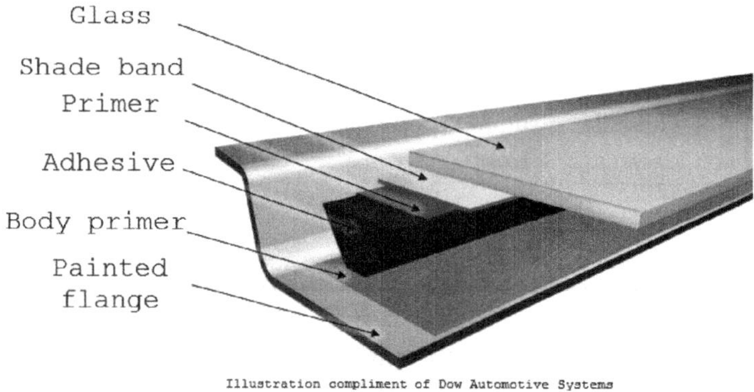



**Figure 8.2** Illustration of layer build-up.

A two-part primer system is preferred to mitigate the inconsistency of the splitting and adhesion of the paint system. (Splitting off may occur when you pull on the urethane, and instead of getting cohesive failure of the urethane to the paint, the paint actually pulls off the body, causing glass to come out in a collision or roll over.) Many applications require two-part priming. This may be due to the assembly plant conditions or the urethane being used. In two-part priming, the primer is applied to the glass as well as the sheet metal, as shown in Figure 8.3.

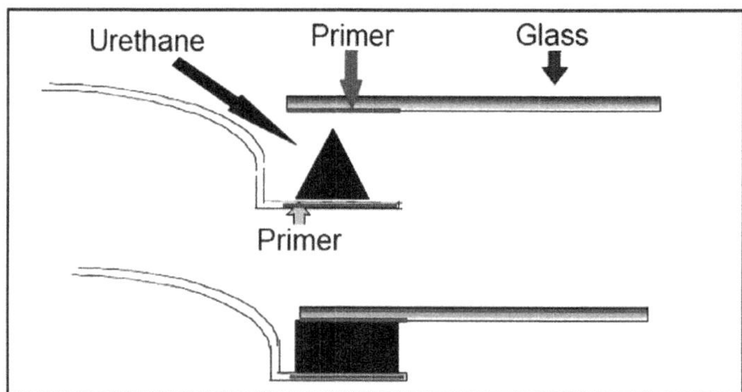



**Figure 8.3** Two-part primer.

# Chapter 9
## Moldings, Encapsulation, and Related Opportunities

Moldings require a minimum of about 50 mm to wrap them properly around the corner without puckering. Less than 50 mm, you will have to miter the corners of the molding, and have the molding pieces welded back together. This becomes costly. A radius ≥ 50 mm will achieve a nicely styled corner, and one that is less costly to manufacture. Figure 9.1 shows a center line section, while Figure 9.2 presents section A-A. Figure 9.3 illustrates a windshield support concept. [9-1]

### 9.1 Molding Types and Materials

When you consider molding type, consider the geometry, especially if there is a corner. The softer the material, the better it will fit in a given space.

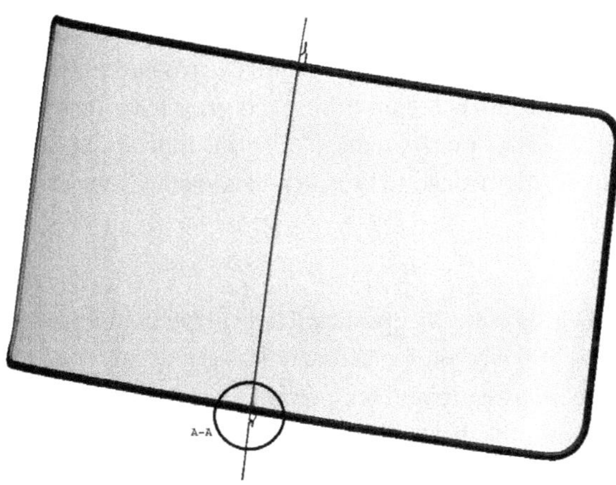

**Figure 9.1** Center line section.

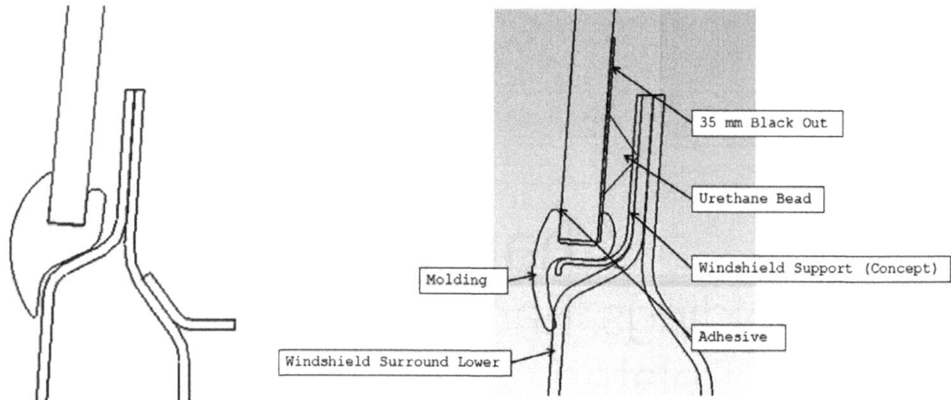


**Figure 9.2** Section A-A.          **Figure 9.3** Windshield support concept.

Choose the type based on your design needs. Each material has its pros and cons depending on profile design. To get softer-durometer ethylene propylene diene monomer (EPDM), you go with a foamed process, but with a thermoplastic elastomer (TPE ) you can just go down in durometer. However, conversely, if you want to increase to the harder durometers with EPDM, you really can only go so far without having to increase size or wall thickness of the profile. When it comes to TPEs, you can simply go to a harder-durometer material all the way into the filled polypropylenes, without having to actually increase profile or wall thickness. This saves in packaging space, material cost, and weight, which is a plus for using TPE (plastics). Additionally, TPEs are fully recyclable because they are thermoplastic and not thermoset materials. Furthermore, whether or not you use TPE or EPDN, for really sharp radii corners you are always better off going to an injection molded corner as a secondary operation.

Many of the following compositions may be good candidates for glazing, depending on the application (e.g., windshield, backlite, fixed quarter window).

Extruded rubber products will differ from molded rubber products based on the process: extruded parts are forced through a die of the required cross-section under pressure of an extruder. Often, extruded products are unvulcanized prior to being extruded, leaving the rubber in a soft and pliable state post-extrusion. If this is indeed the case, the finished extruded products will normally need to be vulcanized before they are rendered usable.

### 9.1.1    Extrusion

The extrusion process begins with the unvulcanized rubber compound being fed into the extruder. Next, the flutes of the revolving screw will begin to carry the rubber forward into the die, with an increase in pressure and temperature occurring as the material gets closer to the die itself. Once it reaches the die, the built-up pressure forces the material through the openings, where it will consequently swell in various degrees based on the material compound and hardness. Because of this tendency toward swelling, many extruded parts require plus or minus tolerances on their cross-sections. During the vulcanization, the extruded rubber will swell or shrink in both its cross-section and its length depending on the type of rubber compound used. After vulcanization, a length of rubber extrusion will tend to be reduced in dimension more in the center of the length than in the ends.

### 9.1.2  Rubber Compression Molding

The rubber compression molding process begins with a sheet of solid, yet soft, rubber that is placed in one half of a preheated mold cavity (usually by hand) of a two-sided mold. The two halves of the mold are fitted together; pressure is applied so the compound within the mold is compressed to fill the mold cavity and form the part. Additional heat applied to the mold forces the rubber to soften further and disperse into all parts of the mold cavity and then heat the rubber to its vulcanization temperature. At a predetermined time-at-temperature, the chemical reaction to change the rubber mixture to a homogenous compound (i.e., vulcanization) is complete, and the finished rubber part is removed from the mold. The mold is cleaned and the process is repeated.

There are many advantages to using compression molding: low tooling cost, minimal waste (because the operator is able to accurately measure the amount of material to fit into the compression mold), and the ability to create intricate products (because high pressure from the compression forces the material to follow the mold to its exact design). Compression molding is ideal for low production volumes of medium-to-large parts as well as for forming bulky parts of low-to-medium production volumes. Rubber compression molding is highly suited for gaskets, seals, and O-rings.

### 9.1.3  Rubber Injection Molding

Rubber injection molding is an ideal process for forming high-volume production rubber parts. This process is well suited for large quantities of small to medium-size parts with complex inserts and close dimensional tolerances. With injection molding, an exact amount of rubber is forced or "injected" into a closed cavity formed by two halves of a steel die that is preheated to the rubber's curing or vulcanizing temperature. Because the injection action is both rapid and high pressure, the die is held in place by the high pressure, thus forming the part cavity, and the die is at temperature, thus accelerating the curing process.

So, short cycle times, excellent dimensional capabilities imparted by the die, and little process waste all come together to make injection molding a low-cost process for high-volume parts. Because the cost of high-precision dies is quite high, this technique is not suited for lower-volume applications.

Rubber injection molding is used widely for many high-volume products for consumer applications (rubber dog bones and door stops), industrial applications (grommets, O-rings), and transportation industry applications (protective boots for vulnerable automotive and truck steering and suspension systems).

## 9.2 Rubber Materials for Extrusion and Molding

### 9.2.1 EPDM

EPDM rubber is a synthetic rubber used in a wide range of applications. It is resistant to heat, UV light, ozone, weather, and water swell. Along with EPDM rubber's excellent heat, ozone, and weather resistance is its good resistance to polar substances, steam, and electrical insulation.

EPDM rubber is an excellent material, often used because of its superior ability to resist degradation due to the sun's UV light and the ozone this generates. EPDM rubber is an affordable material. EPDM "sponge" is often matched with EDPM "dense" to form a resilient seal that can also be fashioned into a "clip," offering a simple, yet effective, attachment mechanism. Various styles of metal carriers are often embedded into the clip to add rigidity and improve fastening capability.

Disadvantages include the fact that EPDM rubber is not resistant to most oils, fuels, solvents, and acids. EPDM rubber exhibits average-to-good resistance to fireproof hydraulic fluids, ketones, hot and cold water, and alkalis. However, EPDM rubber has poor resistance to most oils (petroleum, mineral, and diester lubricants), fuel (gasoline, kerosene), aromatic and aliphatic hydrocarbons, halogenated solvents, and acids.

EPDM rubber is frequently used for weather sealing applications on all vehicles: door, window, trunk, and hood seals. It is also used in glass-run channels, radiator hoses, garden and appliance hoses, tubing, washers, belts, solar panels, gas masks (instead of silicone), and electrical insulation.

### 9.2.2 Nitrile Rubber

Nitrile rubber is a copolymer of butadiene and acrylonitrile. Its use presents some advantages: it is resistant to heat, petroleum-based oils, and fuels over a wide range of temperatures. Specific property requirements are satisfied by adjusting the proportion of each polymer used. Higher acrylonitrile content increases resistance to heat, petroleum-based oils, and fuels, but decreases flexibility in low temperatures. Nitrile rubber's petroleum resistance, combined with its ability to be compounded for service over a wide range of temperatures, make nitrile rubber one of the most widely used elastomers in the seal industry. Nitrile rubber exhibits excellent compression-set resistance and good tear and abrasion resistance.

Disadvantages include the fact that nitrile is not resistant to ozone, weather, or heating aging. Nitrile rubber is used in the automotive and aeronautical industries to make fuel and oil handling hoses, seals, and grommets. It is also used to create molded goods, footwear, adhesives, sealants, sponge, expanded foams, and floor mats. Its resilience or snap makes nitrile rubber a useful material for disposable lab, cleaning, and examination gloves.

## 9.2.3  Neoprene Rubber

Neoprene or polychloroprene is a family of synthetic rubbers produced by polymerization of chloroprene. Neoprene is the DuPont trade name for its brand of polychloroprene.

Neoprene rubber is classified as a general-purpose elastomer with good compression set resistance, good resilience, and the ability to resist abrasion and flex cracking over a wide range of temperatures. Neoprene rubber adheres well to metals for rubber-to-metal bonding applications. Because neoprene rubber is very resistant to Freon and ammonia, it is widely used for sealing refrigeration fluids.

Neoprene rubber is used in a wide variety of applications, such as laptop sleeves, orthopedic braces (wrist, knee, etc.), electrical insulation, liquid and sheet applied elastomeric membranes or flashings, and automotive fan belts. A foamed neoprene containing gas cells is used as an insulation material, most notably in wetsuits and fly-fishing waders. Other foamed neoprene applications include insulation and shock-protection (such as packing).

Neoprene rubber is used in many industrial applications, from face protection masks, waterproof automotive seat covers, to insulation for CPU sockets. As a liquid or sheet, neoprene rubber is used for roof membranes or flashings. As a neoprene-spandex mixture it is used in wheelchair positioning harness manufacturing.

## 9.2.4  HNBR Rubber

HNBR (hydrogenated nitrile butadiene rubber) is a derivative of nitrile rubber that is hydrogenated in solution using precious metal catalysts. During this process, the nitrile groups remain unchanged while the carbon-carbon double bonds in nitrile rubber convert into more stable single bonds.

Similar to other elastomers, HNBR rubber displays good compression set resistance, tear strength, tensile strength, and abrasion resistance, as well as high elasticity. Along with these strengths, HNBR rubber also has good heat-aging resistance and low-temperature tolerances compared to other heat- and oil-resistant elastomers.

With HNBR rubber, as with nitrile, an increase in the acrylonitrile content increases resistance to heat and petroleum-based oils and fuels, but decreases the low-temperature performance. It provides good viscoelastic properties, a wide service temperature range, and resistance to strongly alkaline and aggressive fluids.

With an array of beneficial properties, HNBR rubber continues to find its way into a broad range of applications, especially for the automotive industry. Other properties of HNBR rubber making it highly useful to the automotive industry include good viscoelasticity (property of materials that exhibit both viscous and elastic characteristics when undergoing deformation) when HNBR vulcanizes, a wide service temperature range, resistance to fluids of various chemical compositions, and resistance to strongly alkaline and aggressive fluids.

HNBR rubber exhibits a selection of good properties that usually are not found together in elastomers: heat-aging resistance, oil resistance, and low-temperature flexibility. As the automotive industry requires improvements in performance and reliability, this combination of properties is pushing HNBR rubber to meet more of those requirements.

## 9.2.5  Silicone Rubber

Silicone rubber is a semi-organic elastomer. It resists temperature extremes, compression set, heat aging, ozone, and water swell. It also does well under extreme temperatures while resisting compression set and retaining flexibility.

Silicone rubber does present some disadvantages, including low physical strength and poor resistance to tearing and abrasion. These factors limit its use for static seal applications requiring materials suitable for high friction that are used often in extreme environments.

Compared to conventional rubbers at extreme temperatures, silicone elastomers perform better with greater tensile strength, elongation, tear strength, and compression set resistance, but do not perform as well as some other elastomeric materials. Also, compared to organic rubbers, silicone rubber displays good resistance to the sun's UV light and the ozone this generates, as well as good heat-aging resistance. Unlike organic rubbers, silicone rubber has a low tensile strength, and applications with even low imposed loads must be designed carefully. Silicones are traditionally a more expensive option when compared to EPDM rubbers and thermoplastic elastomers.

Silicone rubber is generally nonreactive, stable, and able to maintain its useful properties through a wide range of temperatures. Due to these properties, and its ease of manufacturing and shaping, silicone rubber is found in a wide variety of products, including automotive applications, home goods (cooking, baking, and food storage products), apparel (undergarments, sportswear, and footwear), electronics, medical devices and implants, and in-home repair and hardware.

## 9.2.6  Polyacrylate

Polyacrylate rubber is a copolymer of ethyl and acrylates. It displays good resistance to petroleum fuels, oils, and grease, even at temperatures up to 302°F (150°C). Also, polyacrylate rubbers exhibit excellent heat-aging properties, such as resisting cracking when exposed to ozone and sunlight. Polyacrylate rubber does not withstand brake fluids, chlorinated hydrocarbons, alcohol, or glycols.

Polyacrylic rubbers are suited to applications such as automotive automatic transmissions and steering systems, which need resistance to petroleum and the ability to withstand high temperatures.

## 9.2.7  Fluoroelastomer

Fluoroelastomer (FKM) is a special-purpose fluorocarbon-based synthetic rubber. Fluoroelastomers exhibit broad chemical resistance and very good performance in high-temperature applications. They also show very good gasoline resistance swelling and good resistance to the sun's UV light and the ozone this generates. Because fluoroelastomers have good gas impermeability, they are ideal for automotive fuel applications and aircraft engine seals.

Fluoroelastomers are compatible with hydrocarbons, and incompatible with ketones and organic acids.

They are generally compatible with hydrocarbons, but incompatible with ketones (acetone) and organic acids (acetic acid).

Fluoroelastomers do present some disadvantages. As a class of materials, they are very expensive, and their use is reserved for highly specialized applications. Common trade names used for these materials include Viton (DuPont) and Technoflon (Solvay Solexis S.p.A.).

When exposed to low temperatures, fluoroelastomers can become quite hard but are still serviceable. Fluoroelastomer compounds are not suited to applications needing flexibility at low temperatures. The performance of fluoroelastomers in aggressive chemicals depends on the nature of the base polymer used and the compounding ingredients used for molding the final products: these compounds are finely tuned to fit the end-use application. Common uses include O-rings, shaft seals, and engine seals.

## 9.3　Thermoplastic Extrusion (Plastics)

Thermoplastic extrusions are normally manufactured in a continuous process in which thermoplastic raw materials (resins) are continuously fed from a resin hopper into a temperature-controlled extruder, where an auger screw propels the fast-melting plastic (under pressure) through a die that forms the material into the desired profile or shape. This shape then enters a cooling medium, usually air or water. It is conveyed through the cooling medium by way of a puller at the end of the extrusion line, which maintains line tension and aids in the overall dimensional stability of the process. This relatively simple process allows for multiple extruders to be linked at the die to provide multi-material profile shapes, where unique material properties can be imparted for the creation of specific applications.

Thermoplastic-extruded components are typically provided in single, co-, or tri-material constructions. Thermoplastic resin materials are normally delivered in granular form (pellets) to the extruder, using suction-feed systems.

Thermoplastic materials can be classified in two primary categories: polar or non-polar, based on specific chemical structure and properties. Traditional polar thermoplastic materials such as PVC (polyvinyl chloride) are not chemically compatible with traditional non-polar counterpart materials such as olefinic blends, including EPR/PP (ethylene propylene rubber/polypropylene).

In recent years, this area of polymer extrusion technology has "exploded" in size and scope of applications over its rubber counterpart. This expansion has been driven by advantages including substantial energy and capitalization savings (from utilization of lower manufacturing temperature requirements), substantially reduced curing (or cooling) times, recyclable offal (process waste and scrap), and easy color matching to end-use applications.

Thermoplastic extrusions are especially well-suited for many transportation and general industry applications. Automotive interior and exterior applications are replete with durable examples of thermoplastic extruded products that are color matched, textured, or laminated to complement the specific style of the vehicle.

The existing uses in home applications that promote thermoplastic extrusions are incalculable. For industrial applications, one only needs to look at grocery store refrigerator seals to see an important use of these versatile materials.

Many thermoplastic materials, owing to their inherent nature, can lose their rebound or elastic properties when overexposed to excessive heat or thermal conditions. This fact excludes these materials from being an optimal choice in various applications, including dynamic door and sunroof seal applications.

## 9.4    Plastic Materials for Extrusion and Molding

### 9.4.1   Thermoplastic Elastomer

TPEs, sometimes referred to as thermoplastic rubbers, are a class of copolymers or a physical mix of polymers (usually a plastic and a rubber) that consist of materials with both thermoplastic and elastomeric properties. While most elastomers are thermosets, thermoplastics are, in contrast, relatively easy to use in manufacturing (e.g., by injection molding).

Thermoplastic elastomers show advantages typical of both rubbery materials and plastic materials. The principal difference between thermoset elastomers and thermoplastic elastomers is the type of crosslinking bond in their structures. In fact, crosslinking is a critical structural factor, which contributes to high elastic properties. The crosslink in thermoset polymers is a covalent bond created during the vulcanization process.

### 9.4.2   Thermoplastic Polyolefin

Thermoplastic Polyolefin (TPO) is a trade name that refers to polymer filler blends usually consisting of some fraction of PP (polypropylene), PE (polyethylene), BCPP (block copolymer polypropylene), rubber, and reinforcing filler. Common fillers include, though are not restricted to, talc, fiberglass, carbon fiber, wollastonite, and MOS (metal oxysulfate). Common rubbers include EPR (ethylene propylene rubber), EPDM, EO (ethylene-octene), EB (ethylene-butadiene), SEBS (styrene-ethylene-butadiene-styrene).

Currently, there is a great variety of commercially available rubbers and BCPPs.

### 9.4.3   Thermoplastic Elastomers Review

Thermoplastic elastomers or thermoplastic rubbers are copolymers of a plastic and a rubber [9-1]. The properties of thermoplastic elastomers reflect the properties of both thermoplastics and elastomers.

There are many commercially available products of TPE elastomeric alloys: Dryflex, Mediprene (Hexpol), Santoprene (ExxonMobile), Geolast (Monsanto), Sarlink (DSM) and

Alcryn (Du Pont). Thermoplastic elastomers have the potential to be recyclable, are easily colored, and provide good batch-to-batch consistency. They allow for improved consistency in raw materials and the finished product, as they need little or no compounding, and no additional reinforcing agents, stabilizers, or cure processes. Therefore, every batch requires the same weight of the same components. Thermoplastic elastomers consume less energy and increase product quality over many other possible material selections. This makes it easier and more economical to control.

Thermoplastic elastomers accept most dye types readily. The plastic component of a thermoplastic elastomer product is reusable and recyclable; the rubber component is not because of its thermoset properties. New technologies are opening up the ability to recycle thermoplastic elastomers.

The disadvantages of thermoplastic elastomers relative to conventional rubber are the relatively high cost of raw materials, general inability to load them with low-cost fillers such as carbon black, poor chemical and heat resistance, high compression set, and low thermal stability.

Thermoplastic elastomers are widely used in a variety of automotive and household appliance sector applications. They are commonly used to make suspension bushings for automotive performance applications because of their greater resistance to deformation when compared to regular rubber bushings.

Thermoplastic elastomers are also finding more and more uses as electrical cable jacket/inner insulation material. Interior trim applications have largely devolved to thermoplastic elastomers vs. the use of more conventional rubbers.

Window main frame components are made using a combination of polypropylene + 15% mineral and UV stabilizer and are tested under 5-year Florida conditions, which are the norm in the industry. Polypropylene is used because of its low cost (a necessity considering the volumes), its resistance to wear and vibrations, its flexibility, and its ability to resist to UV when properly stabilized. Thermal expansion is brought into a reasonable range (similar to unfilled nylons) with the use of 15% mineral filler. A higher percentage of filler would help reduce the expansion further, but it would also increase brittleness.

## 9.5    Encapsulation

Encapsulation is the process of combining the main component (e.g., windshield) with a molded edge or molding, with the ability to add attachment hardware. Figure 9.4 shows an encapsulated quarter window. The advantages are a more uniform aspect and appearance, and convenience at point of assembly. Other advantages include:

- Much improved reduction of stack up tolerances

- Reduction of part numbers needed for the assembly

- A potential reduction and/or elimination of suppliers

- Increased design freedom

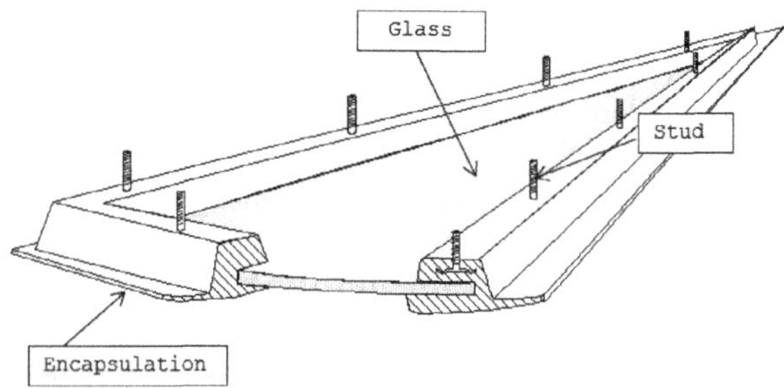


**Figure 9.4** Encapsulated quarter window.

Disadvantages include:

- High initial tooling cost, ~Δ400% increase.

- High service parts and warranty part cost, ~Δ400–800% increase.

- Must replace system, not just component.

- Encapsulation tools are designed for 0.5 mm from maximum glass material condition.

Without adjusting the encapsulation tool, we would not be able to thicken the glass. Maintaining the tool to ensure long tooling life and a quality part is very critical.

Mold release during encapsulation is another key issue. Many suppliers have seen numerous failure modes at plants and OEMs in the past on encapsulated glass, where the bond is with the encapsulations to a body flange. The failure does not occur on every part, nor is this a problem on every type of encapsulation. The problem usually occurs on approximately every 10th part from a run. It can be mitigated sometimes by visual inspection of the parts and then cleaning with alcohol and other cleaners that have been mentioned previously in this book. However, sometimes the mold release surface pollution is not readily apparent visually. EPDMs are different, and although it can be relatively easy to get a short-term bond between the adhesive and EPDM, longer-term adhesion sometimes can be problematic because EPDM continues to "weep," to its surface, chemical contaminants that compromise the adhesion interface.

When you design an extruded section that will become a molded part, you have to consider at least two factors: material type, which dictates the durometer or flexibility, and the section design itself. Keep the section as simple as possible while still able to accomplish its desired outcome. It has to be of a material to bend without puckering and possess a section design to meet its designed expectations. With a single piece that has to follow a radius of approximately 50 mm or less, you will more than likely end up with a puckering condition, as shown in Figure 9.5.

Highlighted below are some reasons that cause a molding to pucker and possible corrective actions.

- Corner joint molding, while not aesthetically pleasing, would enable you to keep the cross-section tight in the radius connecting while linking the two linear elements.

**Figure 9.5** Molding puckering around radius.

- "Key slot," which is a small usually V-shaped slot inserted with a harder durometer material that aids in applying added pressure to the sheet metal surface. The "Key" is a feature used in "Roped in" bus windshields to keep the costs and field repair down to a minimum (see Figure 9.6, Figure 9.7, Figure 9.8, Figure 9.9, Figure 9.10, and Figure 9.11).

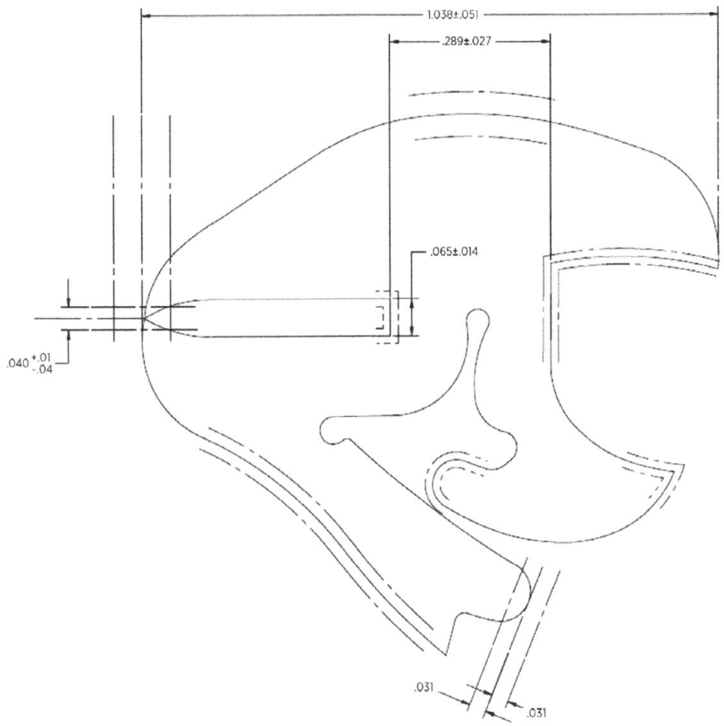

**Figure 9.6** Typical extruded section.

- Another reason you use a key type molding on a window assembly such as a school bus is the molding is not bonded to the glass as passenger cars and trucks.

- When you try to get your molding to do too much, say change direction too quickly as with an accelerated curve, 50 mm or less, this condition will produce gaps and poor contact with the sheet metal.

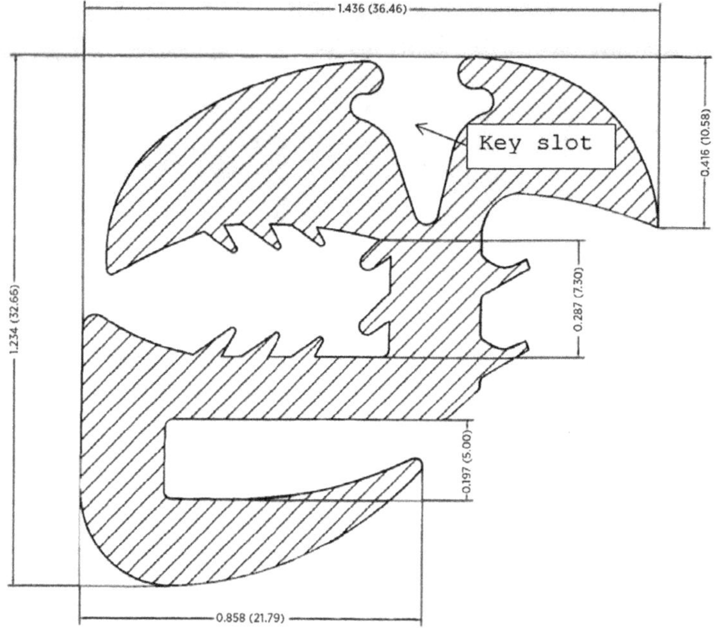


**Figure 9.7** Extruded section with key slot.

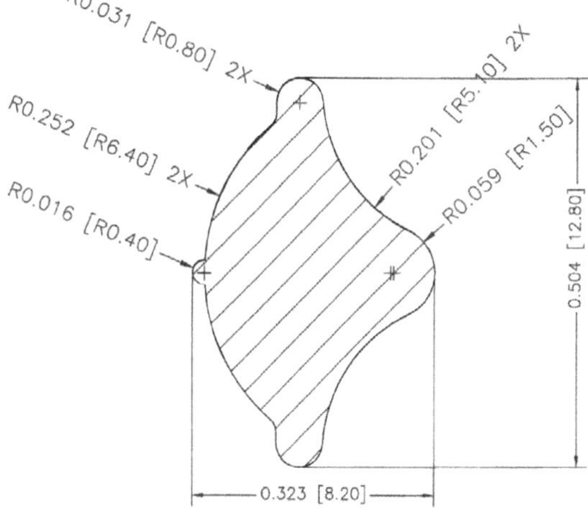


**Figure 9.8** Key slot.

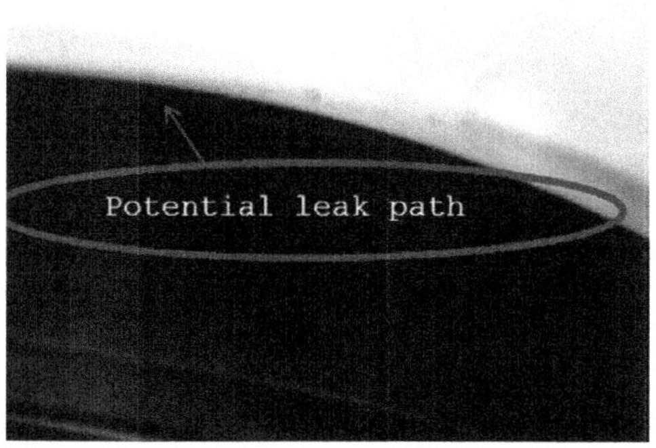


**Figure 9.9** Potential leak path.

**Figure 9.10** Poor contact with
sheet metal.

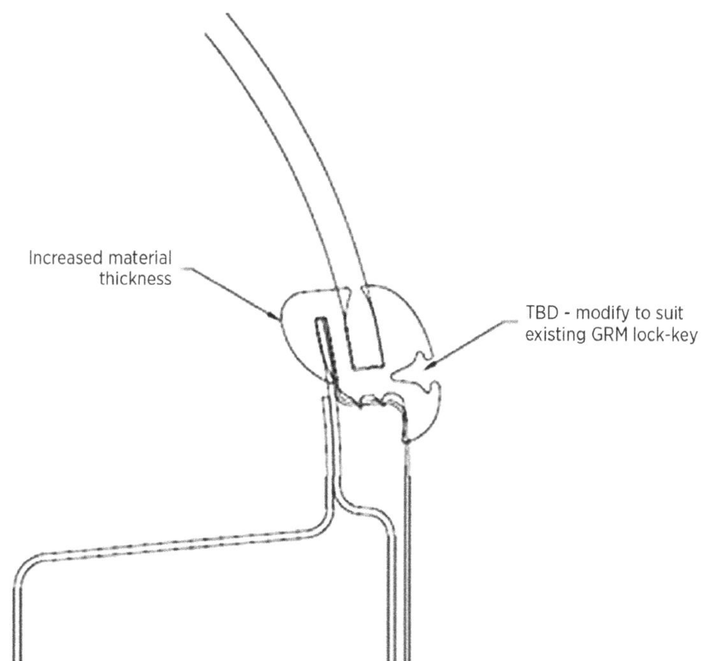


**Figure 9.11** Molding section with key slot in body.

## 9.6    Reference

9-1.    Harper, Charles A. 2002. *Handbook of Plastics, Elastomers, and Composites (4th Edition).*
McGraw-Hill Professional.

# Chapter 10
## Installation and Post Processing

After the urethane has been properly applied to the windshield, it still has to be mounted on the body opening. There are a few ways to achieve this, such as using locating stops, guides, etc. However, locating pins remain the most widely used and offer the most effective way to control surface movement.

## 10.1  Locating Pins and Slots

Locating pins are made from a variety of materials, mostly nylon or a derivative. Pins could be placed across the top edge of glass on the inside surface, as well as on the sheet metal cowl, space permitting. Slots would be pierced in the sheet metal that would receive these pins.

Other methods of holding the glass in place are using setting blocks at the bottom and using tape at multiple locations across the top. Many vehicles use tape across the top at four locations. Most windshields used in bus applications are very heavy, and the installation angle is close to vertical.

## 10.2  Inspection

Many of these inspections occur at the glass manufacturing plant; some take place at the Original Equipment Manufacturer's (OEM's) plant. Between the glass processing plant and final assembly, a few very critical steps are performed. It is this phase that will ultimately determine the quality of the part as well as warranty and potential legal implications.

The items to be inspected are the glass itself, the molding fit, and, if used, encapsulation. Encapsulation tools are designed for 0.5 mm from maximum glass material condition. Without adjusting the encapsulation tool, we would not be able to thicken the glass.

## 10.3   Plant Applied Primer

After the encapsulation has been applied to the component, it still has to be primed for installation. Most windshields come into the plant pre-primed, or primed at the glass supplier. In some cases they are done in the assembly plant.

The major potential problem with plant-level primed glass as opposed to pre-primed glass is operator error. If the operator misses a step in applying the clear, black, or the primer step altogether, the results can be disastrous.

## 10.4   Failure to Prime

If a windshield, for example, was not properly primed, it would take very little force to dislodge it from the cab. This most certainly would cause water leaks into the cab, which would lead to a warranty claim. Worse than that, it could cause very serious injury to the occupant or, in a crash situation, occupant ejection could occur.

## 10.5   Structural Integrity

The windshield installed into the cab opening becomes an integral structural component of the system. No vehicle will pass a roof crush test without it. Always pay critical attention to the weakest, and many times overlooked, process in your system; in this case it is priming.

## 10.6   Quality Control

Quality is such a critical element in the overall design process in glass. As a design engineer, you may think that that you have done due diligence in specifying your part, only to find out that it has failed due to poor edge finish, anomalies is the composition, or poor quality control. When your part fails in the field, the first questions asked of you will be, "Was it to print? How did it match up to the gauge?"

# Chapter 11
## Value-Added Engineering

"Value added" is a term coined by the glass manufacturers to increase customer options at a competitive cost. These features are either added on, or they are built into the product. Many features are typically found on higher-volume or premium models. The following sections briefly discuss some of the value-added items, along with their benefits and relevant technical terms, as well as a presentation of some potential field issues and potential ways to alleviate them.

## 11.1  Infrared Reflecting Glass—Solar Absorbing

### Benefits and uses of Infrared Reflecting (IRR) glass:

- Improves occupant comfort.
- Reduces thermal load on the interior.
- Reduces electrical usage for HVAC system.
- Improves fuel economy/reduces emissions. National Renewable Energy Laboratory (NREL) studies show fuel savings of up to 10 gal, per vehicle, per year, or approximately 2.3 billion gallons in the U.S. annually if the fleet were converted. This equals 47 billion pounds of greenhouse gas per year.
- Ideal for hybrid vehicles.
- Ideal for electric vehicles.

### Technical data and terms:

**Visible Light:**   Radiant energy in the wavelength range of 380 nm to 780 nm with Ill. (Illumination) D65 and CIE 2 observer.

**% Transmittance:**   Percentage of visible light at normal incidence (90° to surface) directly transmitted through the glass.

**% Reflectance Indoors:**    Percentage of visible light at normal incidence directly reflected from the glass back indoors.

**% Reflectance Outdoors:**    Percentage of visible light at normal incidence directly reflected from the glass back outdoors.

**Solar Energy (Direct):**    Radiant energy from the sun having a wavelength range of 300 nm to 2500 nm at ASTM air mass of 1.5.

**% Transmittance:**    Percentage of solar energy at normal incidence (90° to surface) directly transmitted through the glass.

**Reflect % Out:**    Percentage of solar energy at normal incidence directly reflected from the glass back outdoors.

**U-Factor (also called U-Value):**    Air-to-air thermal conductance of 39-in-high glazing and associated air films. Units are Btu/h·ft²·°F.

**UV Light:**    Invisible to the human eye. The wavelength range of UV is 300 to 380 nanometers (nm). Long-term UV exposure can result in fading and deterioration of fabrics and furnishings. Infrared light is also invisible to the human eye. It falls between approx. 790–3000 nm and has a penetrating heat effect. The only portion of the solar spectrum that is visible to the human eye is the visible light range, 380–720 nm.

**Winter-night:**    12.3 mph wind at −0.4°F (−18°C) and 69.8°F (21°C) indoors.

**Summer:**    0 sun, 6.15 mph wind at 89.6°F (32°C) and 75.2°F (24°C), still indoor air.

**Shading Coefficient (SC):**    Fraction of solar heat, direct (300 to 2500 nm) plus indirect (5 to 40 µm), transferred indoors through the glass. For reference, 1/8 in (3.1 mm) clear glass has a value of 1.00 (SC is an older term being replaced by the SHGC).

**SHGC (Solar Heat Gain Coefficient):**    Fraction of solar energy incident on the glazing that is transferred indoors both directly and indirectly through the glazing. The direct gain portion equals the direct solar transmittance, while the indirect is the fraction of the solar energy absorbed to the energy reradiated and convected indoors. No heat gain from warmer outdoor air is included.

**SHGC** = (Direct Solar Trans.) + {[(Indirect Solar Heat Gain) − (Summer U-Value) (89.6°F−75.2°F)]/(248.209 Btu/h·ft²)}.

**Relative Heat Gain (RHG):**    Total net heat gain to the indoors due to both the air-to-air thermal conductance and the solar heat gain. The units are Btu/h·ft².

**RHG** = [(Summer U-Value)(89.6°F−75.2°F) + (Shading Coefficient)(200 Btu/h·ft²)].

## 11.2    Photovoltaic Glass

Photovoltaic (PV) glass is a special glass with integrated solar cells to convert solar energy into electricity. PV is rarely used to provide motive power in transport applications, but is being used increasingly to provide auxiliary power in boats and cars. Some automobiles are fitted with solar-powered air conditioning to limit interior temperatures on hot days.

## Technical data and terms:

**Response Time:**   Approximately 10 µs, virtually instantaneous.

**Framing:**   PU RIM encapsulation possible.

**Durability:**   Operating range, –22 to +230°F (–30 to +110°C), LC durability enhanced by UV-absorbing front protective layer.

**Acoustics:**   Sound attenuation superior to acoustic laminates.

**Solar Properties:**   Governed by base construction and unchanged by switching state. Wide range of performance possible, solar reflective coatings optional. UV Transmission near zero.

**Visible Transmission:**   Governed by construction, 10% to 75% LTa available. Limitless construction variants.

## 11.3   Heated Glass

Heated glass is primarily used to defog/defrost the backlite. However, in some commercial vehicles this technology is being applied to the windshield.

The main reason heated glass has not gained greater acceptance in mainstream automotive windshield applications is cost, closely followed by providing a potential leak path for water intrusion. The heating element is embedded into the PVB; sometimes this produces a delamination.

## 11.4   Laminated Sidelites

The National Highway Safety Testing Agency indicates improved occupant retention with the use of laminated sidelites, which complements side curtain airbags. Ejection statistics are significant:

- 52,897 annual ejections (1995–2003)

- 1% of all crash-involved occupants

- 10,210 annual ejected fatalities

- 32% of all fatalities

- 6124 through side windows

- 10,177 annual rollover fatalities

- 3703 ejected through side windows

## 11.5   Potential Field Issues

### 11.5.1   Mitigating Fogging on Multi-Pane Windows

In a case in the field, side entrance door glass was fogging up due to a poor seal and extremely humid conditions. This generally happens in humid climates when the vehicle is parked for long periods of time.

The solution to this problem is to apply a desiccant between the two panes of glass. The entire assembly then needs to be sealed with silicone.

## 11.5.2  Delamination: An Example

In an example in the field, Glass Company A has reviewed the report of a delamination issue in a location with subtropical climate. The symptom of the issue was identified as "finger delamination." Finger delamination looks like a hand or fingers that cause pockets of air/moisture in the Polyvinyl Butyral (PVB). A possible cause of moisture in the inner layer is environmental moisture that becomes trapped between the glass/encapsulation and the sheet metal opening.

## 11.5.3  Primer and Primer Path Concerns

Both primer and urethane must be applied to bond your glass successfully. Some of the causes why urethane or primer misses the mark on this may be hard to pinpoint. First, has the material been stored properly? Is it applied manually? If it is robotically applied, the cause may be in the programming of the robot. Seek counsel from your urethane supplier before calling the robot manufacturer.

# Chapter 12
## Testing Tables

This book was intended to be a portable guide for the new and experienced designer and engineer. The reader may start at the beginning to understand ceramic engineering or jump to a special area of concern found in the next chapters.

Much value will be derived from the Design Guide Reference Tables presented in Chapter 14. They will provide you and your company a quick way of locating a common challenge, making an educated decision, and knowing the outcome before resources are more fully committed. These guidelines are used by manufacturers worldwide.

Glass is the number one or two warranty claim found in most vehicle manufacturers globally. Being able to properly design it, source it, and have it live up to its product performance promise is a fundamental aspect of one's competitive advantage.

| Table 12.1 Standard Automotive Water Pressure Testing | | | |
|---|---|---|---|
| | (Garden) Hose Test | Car Wash (Pressure Spray) Test | Rain Shower Test |
| Company A | Flow rate = 30 L/min | Flow rate = 2.5 L/min | Rainfall flow rate = 25 mm/h |
| | Pressure = 0.08 MPa | Pressure = 8.3 MPa | Horizontal winds = 30 km/h (19 mph) |
| | Water temperature = 140°F (60°C) | Water temperature = 176°F (80°C) | Vehicle stationary at any position up to 20° from horizontal (during or after test) |
| | Flare angle at nozzle = 25° | Flare angle at nozzle = 25° | Vehicle at speed of 135 km/h (84 mph) |
| | Angle of impingement on window = 30° and 90° | Angle of impingement on window = 30° and 90° | |
| | Distance (window to nozzle) = 305 mm | Distance (window to nozzle) = 80 cm | |
| | During test, vehicle stationary and horizontal | During test, vehicle stationary and horizontal | |
| | After test, vehicle stationary and at any angle up to 20° from horizontal | After test, vehicle stationary and at any angle up to 20° from horizontal | |
| Company B | Flow rate = 30 L/min | Not tested | Flow rate = 1081 L/min |
| | Hose diameter = 13 mm | | Test time = 15 min |
| | Hose positioned 100 cm from window | | 32 nozzles for even distribution over surface of vehicle |
| | Sweep rate = 100 mm/s | | Nozzles generally directed perpendicular to |
| | 3 passes around seal | | |
| Company C | Flow rate: Defined as a length of hose (19 mm inner diameter) with the last 1.0 m directed vertically upward must have a spout height between 20 and 22 cm | Flow rate = 760 L/h | Discharge pressure = 200 kPa |
| | Hose diameter = 19 mm | Pressure = 5.0 MPa | Spray spread uniformly over surface |
| | Hose position = 10 cm, perpendicular to window | Angle of impingement on window = perpendicular to window | Spray directed generally perpendicular to surface |
| | Sweep rate = 250 mm/s | Distance (window to nozzle) = 150 cm | Nozzle 80 to 100 cm from surface |
| | 2 passes around seal | | Test time = 10 min |

| Table 12.2 Transit Bus Window Benchmarking | | | |
| --- | --- | --- | --- |
| | **Company A** | **Company B** | **Company C** |
| Slide Efforts (Avg.) | 13 lb | 19 lb | 22 lb |
| Latch Efforts (Avg.) | 3.8 lb | 6 lb | 11 lb |
| DLO Size | 23.5-in × 25.5-in | 25-in × 30-in | 21-in × 24.5-in |
| Glass Thickness | 3.2 mm | 3.2 mm | 3.2 mm |
| Number of Components | 22 components<br>+ 16 screws<br>——————————<br>8 aluminum channels<br>2 grommets<br>2 latches<br>2 springs<br>2 seals<br>2 glass<br>2 stops<br>Sealant | 24 components<br>+ 8 screws<br>——————————<br>6 molded pieces<br>7 aluminum channels<br>2 latches<br>2 springs<br>2 latch covers<br>2 foam strips<br>3 EPDM extrusions<br>Primers<br>Adhesive<br>Sealant | 22 components<br>+ 10 screws<br>——————————<br>9 aluminum channels<br>2 grommets<br>2 latches<br>2 springs<br>5 extrusions<br>2 glass |

| Table 12.3 Coding System for Model-Numbers Required by ANSI Z 26.1—1996 | | | | |
|---|---|---|---|---|
| 1st digit—material | 2nd digit—color/interlayer construction | 3rd and 4th digit— nominal thickness | 5th digit—color/incul. glazing only | 6th/8th resp. 6th/7th digit—special treatment |
| 1 laminated windscreens (AS1) | 1 clear | 44 4.4 mm | 1 different colors e.g., green/clear | -1 0.76 acoustic PVB or 0.50 acoustic PVB |
| 2 laminated glazing (AS2) | 2 green | 32 3.2 mm | 2 identical colors e.g., green/green | -2 wedged interlayer |
| 3 toughened glazing (AS2) | 3 bronze | 17 17 mm insul. glass | | -3 0.75-0.38 PVB |
| 4 laminated glazing (AS3) | 4 blue | 19 19 mm insul. glass | | -4 3M IR-interlayer 0.05 mm |
| 5 toughened glazing (AS3) | 5 grey | 50 5.0 mm | | -5 0.38-0.05-(0.76 dark tinted) |
| 6 semi-toughened laminated glazing (AS2 or AS3) | 6 coated/printed glass | | | -6 0.38 PVB dark tinted/opaque |
| 7 insulating glazing (AS2) | 7 IR-interlayer/ clear glass 0.38-0.05-0.38 | | | -7 0.76 PVB dark tinted/opaque (likewise 0.38-0.38 PVB) |
| 8 intruder resistance glazing (AS2) | 8 IR-interlayer/ clear or tinted glass 0.38-0.05-0.76 or 0.38-0.76-0.05-0.38 | | | -8 black printed clear glass |
| 9 insulating glazing (AS3) (AS2/AS2/3) | 9 IR-interlayer/ clear or tinted glass 0.38-0.76-0.05-0.38 | | | -9 black printed green glass |
| 0 insulating glazing(AS3) (AS3/AS2) | 0 IR-interlayer/ clear glass 0.38-0.05 (0.025)-0.50 | | | -10 black printed Sundym/Venus/ Galaxsee glass |
| A n.n. | A opaque: white | | | -11 0.76 solar control PVB or 0.38 standard-0.76 solar control PVB |
| B n.n. | B n.n. | | | -12 2 times 0.76 PVB dark tinted = 1.52 mm |
| C n.n. | C n.n. | | | -13 insulating glass with 13 mm gap |
| D n.n. | D n.n. | | | -14 n.n. |
| E n.n. | E n.n. | | | -15 n.n. |
| Examples | | | | |
| M1250 green laminated windscreens, 5.0 mm | M 71192-13 insulating glass (AS2), 19mm clear/clear and 13 mm gap | M 1650-1 5.0 mm laminated windscreens with 0.76 mm acoustic PVB | M 6250 semi-toughened green laminated glass 5.0 mm | |

| ANSI Test No | Test | Size mm*mm | Number pc. | Conditions | Results | Lam AS 1 | Lam AS 2 | Tem AS 2 | MGU AS 2 | Lam AS 3 | Tem AS 3 | Remarks |
|---|---|---|---|---|---|---|---|---|---|---|---|---|
| 1 | Light stability | 305*305, flat | 3 | 100 h UV irradiation + 3 min immersed in water 66°C and after that keeping at 100°C for 10 min | Light transmission >70% and at least 95% of the original LT + no bubbles or delamination in the irradiated portion | X | X | X | X | X | | Additional procedure for laminated only |
| 2 | Light transmission | | | Measurement with illuminant A | Light transmission >70% | X | X | X | X | | | Specimens from test 1 |
| 3 | Humidity test | 305*305, flat | 3 | 2 weeks at 52°C and 100% rel. h. | No delamination, no bubbles to a depth of more than 6.35 mm | X | X | | | X | | |
| 4 | Boil test | 305*305, flat | 3 | 3 min 66°C, then 2 h boiling water | No delamination, no bubbles to a depth of more than 13 mm | X | X | | | X | | |
| 6 | Impact, ball (3.05 m) | 305*305, flat | 12 | 227 g ball, 3.05 m at 21°C–29°C on the outer surface | Not more than 2 samples may break | | | X | | | X | No re-test allowed |
| 7 | Fracture test | Largest part representing a model (M-number) | 6 | Break point shall be 25 mm inboard of the edge at the midpoint of the longest edge of the specimen | No individual fragment > 4.25 g | | | X | | | X | 3.2 mm/530 mm$^2$/23 mm*<br>3.6 mm/470 mm$^2$/22 mm*<br>4.0 mm/425 mm$^2$/21 mm*<br>5.0 mm/340 mm$^2$/18 mm* |
| 8 | Impact, shot bag | 305*305, flat | 5 | 4.99 kg shot bag, 2.44 m at 21°C–29°C on the inner surface | Not more than 1 sample may break | | | X | | | X | |
| 9 | Impact, dart | 305*305, flat | 5 | 200 g dart, 9.14 m at 21°C–29°C on the outer surface | Not more than 1 sample may break | X | X | | | X | | Dart may not penetrate the specimen |
| 12 | Impact, ball (9.14 m) | 305*305, flat | 12 | 227 g ball, 9.14 m at 21°C–29°C on the outer surface | Not more than 2 samples may break, opposite to the point of impact small fragments may leave in an area of not more than 5 × 5 mm$^2$. Total separation of glass shall not exceed 1935 mm$^2$ on either side. | X | X | | | X | | |
| 15 | Deviation and distortion | 305*305, flat and (curved) | 10 (3) | Examination from a distance of 7.62 m, sample normal to the surface | Apply ECE-test method | X | | | | | | Requirements of ECE |
| 18 | Abrasion resistance | 102*102, flat | 3 | Taber 1000 cycles | Haze <2% | X | X | X | | | | Annual test not necessary |
| 26 | Penetration resistance | 305*305, flat | 10 | 2.26 kg ball, 3.66 m at 21°C–29°C on the inner surface | Not more than 2 samples may fail | X | | | | | | |

Lam    Laminated safety glass
Tem    Tempered or toughened safety glass
MGU    Multiple glazed unit/insulating glass
AS 1    For all locations in the car, windscreen included
AS 2    For all locations in the car, windscreen excluded
AS 3    For locations not requisite for driving visibility, e.g. roof lights
* Allowable area of individual fragments respectively length of the edge of a resulting square

## Table 12.5  Sample Design, Validation, Planning, and Responsibility Matrix

| Number | Task | Accountable | Support | Int'l Approval Required? |
|---|---|---|---|---|
| 1 | Maintenance and tracking of detailed timing plan | Supplier | International (identify major milestones) | Yes |
| 2 | Identification of Design Static Audit Objectives (DSAO) and customer quality metrics | OEM plant quality | International | Yes |
| 3 | DFMEA | Supplier | OEM | Yes |
| 4 | DVP&R tracking/ updating | Supplier | N/A | Yes |
| 5 | Identification, resolution & management of open issues (report weekly) | Supplier | OEM | Yes |
| 6 | Window industrial design/ styling (if necessary) | Supplier | OEM | Yes |
| 7 | Window A-surface creation (if necessary) | Supplier | N/A | Yes |
| 8 | FEA and other analysis | Supplier | OEM | Yes |
| 9 | System Integration Engineering (bus body design, interior trim design, electrical harness design, fastener selection) | OEM | Supplier | N/A |
| 10 | Creation of top level window assembly drawings | Supplier | OEM | Yes |
| 11 | Creation/ release of new BOM and installation diagrams | OEM | N/A | N/A |
| 12 | Durability and BSR Testing | OEM | Supplier | N/A |
| 13 | Development of vibration drive data | OEM | Supplier | N/A |
| 14 | Balance of testing | Supplier | N/A | Yes |
| 15 | Window assembly PFMEA | Supplier | OEM | Yes |
| 16 | Window installation PFMEA | OEM | Supplier | N/A |

| Table 12.6 | Legal Requirements, Test Description, and Accepted Criteria | |
| --- | --- | --- |
| **Requirement** | **Test Description** | **Acceptance Criteria** |
| Spectral Transmission Test | Fogging—Test per SAE J1396 | TBD |
| Environment | Effects of Thermal Expansion (CLTE)—Soak part at 230ºF (110ºC) surface temperature for 4 hours in the infrared oven. Measure gaps at temperature. | Design gaps remain within tolerances after cycle. |
| Simulated weathering of glazing materials | Internal Spec#, Test per SAE J 2081, SAE J 1907, SAE J 1396, SAE J 674, SAE J 673, SAE J 216, SAE J100 and ANSI Z26.1 | TBD |
| Peel Test | Per SAE J 1907 | TBD |
| Glass  Bonding | Test per Internal Spec# | TBD |
| Vehicle Acoustics Test | *Squeak and Rattle—Shake assembly | No squeaks or rattles on assembly |
| Custom | Packaging Durability—Shake packed parts | No damage after packaging shaker test |
| Weight | Assembly Weight | xxxxxx kg |
| FMVSS 302 | Flammability | Burn rate less than or equal to requirement |
| FEA Durability | Durability, Modal and Natural Frequency Analysis | The components must not sustain any fracture or breakage and must not have first vibrational resonant modes below 18 Hz nor between 28 Hz and 45 Hz. |

| Table 12.7 Legal Requirements, Test Description, and Accepted Criteria | | | |
|---|---|---|---|
| **Requirement** | **Test / Description** | **Acceptance Criteria** | **Test Responsibility (OEM/Supplier)** |
| Light Stability | ANSI/SAE Z26-1 Sec 5.2 | Should not be affected by exposure to sunlight over an extended period of time. | Supplier |
| Humidity Test | ANSI/SAE Z26-1 Sec 5.3 | No separation of materials shall develop. | Supplier |
| Air and Water Leaks | Internal Spec | Must remain fully functional. | OEM |
| Impact Dart (10 ft) | ANSI/SAE Z26-1 Sec 5.6 | Acceptable impact resistance. | OEM |
| Thermal Shock | SAE J1455 | Must remain fully functional. | OEM |
| Abrasion Resistance | ANSI/SAE Z26-1 Sec 5.18 | Mean percentage of light scattered shall not exceed 2.0%. | Supplier |
| Weathering Test | PER Internal Spec Test per SAE J 2081, SAE J 1907, SAE J 1396, SAE J 674, SAE J 673, SAE J 216, SAE J100 and ANSI Z26.1 | Successfully withstand weathering over an extended period of time. | Supplier |
| Glass Bonding | Internal Spec | TBD | Supplier |
| Durability | Shaker Test | Withstand simulated 12 years or 250,000 miles of use and still be fully functional. | OEM |
| FEA Durability | Durability, Modal and Natural Frequency Analysis | The components must not sustain any fracture or breakage and must not have first vibrational resonant modes below 18 Hz nor between 28 Hz and 45 Hz. | Supplier |
| Dimensional Conformance | SAE J673 | Must conform to SAE standards. | Supplier |

| Table 12.8 Failure Mode Effects and Analysis | | | | | |
| Process Intent | | Process Assessment | | | |
| Process Function | Require-ment(s) | Potential Failure Mode(s) | Potential Effect(s) of Failure Mode | Rank | Potential Cause(s)/ Mechanism(s) of Failure Mode |
|---|---|---|---|---|---|
| Windshield installation | Install windshield | Windshield fragile (brittle or breaking) chips, scratches | Detachment of glass | 9 | Glass with primer expired |
| | | | | 9 | Dirty glass |
| | | | | 9 | Glass without primer |
| | | | | 9 | Glass with incorrect primer |
| | | | | 9 | Primer exceeds expiration date |
| | | | | 9 | Primer not meeting specs |
| | | | | 9 | Primer not applied properly from supplier |
| | | | | 3 | Urethane in disrepair |
| | | | Water filtration | 6 | Not properly installed |
| | | | | 6 | Wavy glass frame |
| | | Excess urethane around windshield | Visual defect | 3 | Excess pressure on the glass when mounted |
| | | | | 3 | Glass bad position on the cabin |
| | | Poor urethane adhesion | Water and air leaks, does not meet gov spec for glass retention | 6 | Incorrect sealer specification |
| | | | Detachment of glass | 9 | Incorrect storage of primer |
| | | Scratched glass | Visual defect | 3 | Glass working on the rack |
| | | Broken glass | Change glass, paint damage possible rework | 6 | Wavy glass frame |
| | | | | 6 | Excess pressure on the glass when mounted |
| | | Primed surface not properly cleaned/ prepped | Contaminated surface/lack of adhesion | 9 | Surface not cleaned properly |
| | | Insufficient urethane applied to windshield | Improper adhesion —does not meet design specs | 9 | Windshield not positioned properly |
| | | Mis-installed windshield into the cab opening | Water leaks/ glass breakage | 9 | Insufficient pressure applied to windshield or loading |
| | | Scratches/ damage to preprimed surface | Lack of adhesion to windshield | 9 | Operator error/ improper handling of windshield |
| | | | | 9 | Robotic applicator removing primer from glass surface |

(continued)

| Table 12.8 | Failure Mode Effects and Analysis (continued) | |
|---|---|---|
| **Rank** | **Effect** | **Effect on Product** |
| 10 | Failure to Meet Safety and/or Regulatory Requirements | Potential failure mode affects safe vehicle operation and/or involves non-compliance with government regulation without warning. |
| 9 | | Potential failure mode affects safe vehicle operation and/or involves non-compliance with government regulation with warning. |
| 8 | Loss or Degradation of Primary Function | Potential failure mode may result in loss of primary function of the product (vehicle inoperable, does not affect safe vehicle operation). |
| 7 | | Potential failure mode may result in a degradation of primary function (vehicle operable, but at a reduced level of performance). |
| 6 | Loss or Degradation of Secondary Function | Loss of secondary function (vehicle operable, but comfort/convenience functions inoperable). |
| 5 | | Degradation of secondary function (vehicle operable, but comfort/convenience functions at reduced level of performance). |
| 4 | Annoyance | Appearance or audible noise, vehicle operable, item does not conform and is noticed by most customers (> 75%). |
| 3 | | Appearance or audible noise, vehicle operable, item does not conform and is noticed by many customers (> 50%). |
| 2 | | Appearance or audible noise, vehicle operable, item does not conform and is noticed by discriminating customers (< 25%). |
| 1 | None | No discernible effect. |

| Table 12.9   How to Conduct Your Design Review | |
|---|---|
| **Items to Address:** | |
| **Describe the Problem:** | Windshield is cracking in low temperatures. |
| **Present Your Containment Plan:** | Defective part/process pulled from service or part manufactured in another facility. |
| **Verify Root Cause:** | Incorrect manufacturing method. |
| | Poor design practices. |
| **Describe Your Interim Corrective Action:** | Change tooling process. |
| **Validate Plan:** | Parts on vehicles. |
| | Prototypes being tested. |
| **Define a Permanent Corrective Action:** | Plant audit. |
| | Went to a new manufacturing method (i.e., Doublet vs. Singlet). |
| **If Necessary Provide Timing and Cost Impact:** | Supplier X states that they can have prototypes available in 12 weeks. Changes will result in $1000 tooling and $1.00 piece price increase. |
| **Is Support Required?** | |

# Chapter 13
## Common Glazing Terms

These terms have been accumulated over thirty years through meetings with suppliers and my ceramics engineering studies. Most glass-related terminology is used only within the small glass and glazing community. Examples are the terms lite, sidelite, backlite, etc. I have included examples where needed.

**Adhesion**   The attraction between two surfaces, as between coatings, adhesives to substrates; it is an important factor in the durability of a bonded windshield. Example: Proper adhesion will determine if your vehicle passes specific government mandated tests (e.g., roof crush), mitigates the opportunity for occupant ejection, and keeps the cabin free from water intrusion.

**Adsorption**   The process by which a substance, usually a solid, attracts and retains on its surface the molecules of another substance. Example: Urethane that is exposed to extreme changes in temperature and humidity would become unstable.

**Angle of deviation**   The angle through which a ray of light is deviated by a refracting or reflecting surface, or a prism; the angle between an incident ray and the refracted or reflected ray. Example: Light as it passes through a curved piece of glass experiences an angle of deviation as it is influenced by the effected surfaces.

**Angle of incidence**   The angle formed between a ray of light striking a surface and the normal to that surface at the point of incidence.

**Angle of reflection**   The angle formed between the normal to a surface and the reflected ray. This angle lies in a common plane with the angle of incidence and is equal to it.

**Annealing**   The process of heating and slowly cooling a solid material such as glass. Example: Windshield processing.

**Annealing furnace**   An oven or furnace that possesses the design requirements and heat control necessary to anneal glass for the optical industry.

**Anti-reflection coating**   A thin layer of material applied to a lite surface to reduce the amount of reflected energy.

**A-Pillar**   Side of the window that wraps around the driver or passenger part of the cab or cockpit.

**Backlite**   Rear window in a vehicle.

**Blank**   A piece of glass formed roughly by molding or cutting into the approximate shape and size of the finished part.

**Borosilicated glass**   A strong, heat-resistant glass that contains a minimum of 5% boric oxide.

**Brewster's angle**   For light incident on a plane boundary between two regions having different refractive indices, the angle of incidence at which the reflectance is zero for light that has its electrical field vector in the plane defined by the direction of propagation and the normal to the surface. For propagation from medium 1 to medium 2, Brewster's angle is given as arctan.

**Center post or C-Pillar**   A sheet metal structural element that gives support to the body and aids in overall vehicular stiffness.

**Chip**   A localized fracture at the end of a cleaved optical fiber or on a glass surface.

**Coefficient of thermal expansion (COE)**   A numerical representation of the rate at which a material will exhibit dimensional changes as a direct result of changes in temperature.

**Cold coating**   A method of applying antireflection coatings to optics that avoids the elevated temperatures normally used. A cold coating will not be as durable as a normal coating.

**Component**   A constituent part. It may consist of two or more parts laminated together, or with near and approximately matching surfaces.

**Concave**   Describing a hollow curved surface; curved inward.

**Converging lite**   Also known as convergent lite or convex lite. A lite that converges an incident bundle of rays to a focus.

**Convex**   Denoting a spherically shaped surface; curved outward.

**Corner reflector**   Also known as a corner-cube prism. A prism having three mutually perpendicular surfaces and a hypotenuse face. Light entering through the hypotenuse is reflected by each of the three surfaces in turn and will emerge through the hypotenuse face parallel to the entering beam. The prism thus returns entering beams to the source. Also may be constructed from three first-surface mirrors.

**Curvature**   The measure of departure from a flat surface, as applied to lites; the reciprocal of radius. Applies to any surface, including lites, mirrors, and image surfaces.

**Curvilinear distortion**   A lite aberration in which the focal length varies radially outward from the center of the field. It has the effect of making a straight tangential line in the object appear curved in the image, either convex outward (barrel distortion) or concave outward (pin-cushion distortion). Straight radial lines remain straight in the image.

**Depth of Bend**   Laying a flat bar between the A-pillar and Center post; if none is present, between the opposite A-pillar, then measuring from the flat bar to the surface to find the greatest deviation.This is the Verticle Depth of Chord (Vdoc).

**DFMEA (Design Failure Modes and Effects Analysis)**   A copy of the Design Failure Modes and Effects Analysis (DFMEA) is reviewed and signed-off by supplier and customer. If customer is design responsible, customer will generally share this document with the supplier.

**Design Requirements Document (DRD)**   Covers the complete engineering scope, use, requirements, testing, and life cycle of the part.

**Diamond cutting tool**   A tool made by embedding small particles of diamond in the working edge. In the optical field, the most commonly used diamond tools are glass saws, cup-shaped tools for curve generators, and diamond mills for forming prisms and flat glass plates. Diamond milling is much quicker than the older usage of iron tools with loose emery or carborundum as an abrasive; moreover, it leaves the glass so smooth as to be almost ready for polishing. Diamond tools do not make deep fissures as loose abrasives do.

**Distortion**   A general term referring to the situation in which an image is not a true-to-scale reproduction of an object. There are many types of distortion: anamorphic distortion, curvilinear distortion, keystone distortion, panoramic distortion, perspective distortion, radial distortion, stereoscopic distortion, tangential distortion, wide-angle distortion.

**Diverging lite**   Also known as divergent lite, negative lite, concave lite, or dispersive lite. A lite that causes parallel light rays to spread out. The lite surfaces may be plano-concave, double-concave, or concavo-convex. The edge of a diverging lite is always thicker than the center.

**Double-concave lite**   A diverging lite with both surfaces concave.

**Double-convex lite**   A converging lite with both surfaces convex.

**DVP&R (Design, Verification, Plan and Report)**   The result of this process is a series of documents gathered in one specific location (a binder or electronically) called the "PPAP Package." The PPAP package is a series of documents that need a formal approval by the supplier and customer.

**Edge**   The flat lite or prism surface on every piece of glass. Example: Edges are numbered from 1 to 4. Edges are customer defined. Prices increase roughly $3.00 per edge type and are based upon vehicle position.

**Epoxy**   Common name for a variety of adhesives used for lite bonding.

**Failure Modes and Effects Analysis (FMEA)**   Failure Modes and Effects Analysis is a step-by-step approach for identifying all possible failures in a design, manufacturing, or assembly process, or a product or service.

**Flat blank**   A piece of glass having a planar or flat surface on each side.

**Fluoride glass**   Optical glass containing zirconium fluoride that results in special characteristics such as improved transmission.

**Fringe**   An interference band such as Newton's ring.

**Frit Pattern**   The black opaque band, with or without a dot matrix pattern, for which the primary use is to hide exposed trim from the viewer and protect the molding from UV light.

**Fused silica**   Glass consisting of almost pure silicon dioxide ($SiO_2$). Also called vitreous silica. Frequently used in optical fibers and windows.

**Glass**   A noncrystalline, inorganic mixture of various metallic oxides fused by heating with glassifiers such as silica, or boric or phosphoric oxides. A common window is a mixture of soda, lime, and sand, melted, cast, and rolled into shape. Most glasses are transparent in the visible spectrum and up to about 2.5 µm in the infrared, but some such as natural obsidian are opaque; these are, nevertheless, useful as mirror blanks. Traces of some elements such as cobalt, copper, and gold are capable of producing a strong coloration in glass. Tempered glass has a high degree of internal strain, caused by rapid cooling, which gives it increased mechanical strength.

**Glass annealing furnace**   A furnace, generally electrically heated, with a control system capable of following a cam by which the temperature can be made to go through a definite cycle over a period of days, or even weeks, if the glass is massive. Glass has a softening temperature above which it can be easily bent or molded to shape, and an annealing temperature below which the properties of the glass remain fixed down to room temperature. Between these two points is the annealing range in which it is essential to cool the glass very slowly to remove all strain and to drive the refractive index up to its stable maximum value.

**Grinding**   The process in the manufacture of an optical system that gives it the required geometric shape. Grinding also relieves edge stress.

**Grinding and polishing machinery**   Machinery used to grind and finish a component, such as a lite or prism, to a desired precision. Usually such machines carry a cup-shaped or flat tool (lap) into close contact with the part. An abrasive or polishing material is placed between the tool and the part, and the lap is moved in a nonrepetitive series of movements. Driving mechanisms for aspheric shaping are complicated; for spherical and flat parts, they are relatively simple.

**Grinding tool**   A tool of cast iron or another suitable medium such a belt or grinding wheel that uses a silicon carbide, aluminum oxide, or emery for grinding optical surfaces.

**Ground glass**   A plate of glass in which a face has been frosted by grinding or etching. It diffuses light by scattering in directions close to the incoming beam, but the light falls off rapidly at larger angles out from this direction.

**Index of refraction**   The ratio of the velocity of light in a vacuum to the velocity of light in a refractive material for a given wavelength.

**Indium Tin Oxide (ITO)**   A material widely used as a transparent conductive coating.

**Interferometry**   The study and utilization of interference phenomena based on the wave properties of light.

**Internal transmittance**   The ratio of the radiant power transmitted to the second surface of a medium to the corresponding radiant power that has just passed through the first surface. Internal transmittance does not denote the effects of surface losses and inter-reflection between the two surfaces. Examples include sunlight, and car headlights passing through the windshield.

**Jig**   A device to hold and locate a work piece as it guides, controls, or limits a cutting tool.

**Lite**   A transparent component consisting of one or more pieces of glass with surfaces so curved (usually spherical) that they serve to converge or diverge the transmitted rays from an object, thus forming a real or virtual image of that object. Every lite acts like a lite.

**Metallic coating**   A thin layer of metal deposited on the surface of the glass. The film may serve as a reflector, beamsplitter, neutral density filter, or electromagnetic interference filter.

**Micron (μm)**   A unit of length in the metric system equal to one millionth of a meter ($10^{-6}$ m). Also called micrometer.

**Millidiopter**   A unit of metric measure equal to 0.001 D. The power of a lite in millidiopters is the reciprocal of its focal length in kilometers.

**Milling**   An automatic surface-generating process involving the abrasion of glass by a diamond-charged wheel.

**Molded blank**   A blank whose basic surface curves are attained by heating and forming a given weight of raw glass; a rough glass blank resembling the finished lite in size and shape. After molding, a precision lite blank must be fine-annealed to eliminate strain and to restore the refractive index to its stable maximum value.

**Multilayer coating**   A blank whose basic surface curves are attained by heating and forming a given weight of raw glass; a rough glass blank resembling the finished lite in size and shape. After molding, a precision lite blank must be fine-annealed to eliminate strain and to restore the refractive index to its stable maximum value.

**Nanometer (nm)**   A unit of length in the metric system equal to $10^{-9}$ meters. It formerly was called a milli-micron.

**Normal**   An axis that forms right angles with a surface or with other lines. The normal is used to determine incident, reflective, and refractive angles.

**Normal incidence**   Light striking a surface at an angle perpendicular to the surface.

**Optical Axis (OA)**   The line passing through both the centers of curvatures of the optical surfaces of a lite; the optical centerline for all the centers of a lite system. Not to be confused with optic axis.

**Optical Density (OD)**   A measure of the transmittance through an optical medium. Optical density equals the log to the base 10 of the reciprocal of the transmittance.

**Optical flat**   A piece of glass having one or both surfaces carefully ground and polished plano, generally flat to less than a tenth of a wavelength.

**Optical surface**   A reflecting or refracting surface contained within an optical system.

**Optics**   That branch of physical science concerned with vision and certain phenomena of electromagnetic radiation in the wavelength range extending from the vacuum ultraviolet at about 40 nm to the far-infrared at 1 mm.

**Peak wavelength**   The wavelength at which the radiant intensity of a source is maximum.

**Perspective distortion**   The distortion that is the result of viewing a print from a point other than the center of perspective. The center of perspective is that viewpoint at which the angular sub tenses of points in the picture are identical with the angular sub tenses of the original points in the scene, at the camera lite. Viewing a picture from too far away results in an excessive magnification of near points compared with distant points.

**Plane**   A surface that has no curvature; a surface that is perfectly flat.

**Polarization**   With respect to light radiation, the restriction of the vibrations of the magnetic or electric field vector to a single plane. In a beam of electromagnetic radiation, the polarization direction is the direction of the electric field vector (with no distinction between positive and negative as the field oscillates back and forth). The polarization vector is always in the plane at right angles to the beam direction. Near some given stationary point in space, the polarization direction in the beam can vary at random (unpolarized beam), can remain constant (plane-polarized beam), or can have two coherent plane-polarized elements whose polarization directions make a right angle. In the latter case, depending on the amplitude of the two waves and their relative phase, the combined electric vector traces out an ellipse, and the wave is said to be elliptically polarized. Elliptical and plane polarizations can be converted into each other by means of bi-refringent optical systems.

**Polishing**   The optical process, following grinding, that puts a highly finished, smooth, and apparently amorphous surface on a lite or a mirror.

**PFMEA**   A copy of the Process Failure Mode and Effects Analysis (PFMEA) is reviewed and signed-off by supplier and customer. The PFMEA follows the Process Flow steps, and indicates "what could go wrong" during the fabrication and assembly of each component.

**PPAP**   The PPAP process is designed to demonstrate that the component supplier has developed their design and production process to meet the client's requirements, minimizing the risk of failure by effective use of APQP.

**Pressing**   A blank having basic surface curves attained by forming heat-softened glass that is pressed in a mold.

**Protective coating**   A film applied to a coated or uncoated optical surface primarily for protecting this surface from mechanical abrasion, from chemical corrosion, or from both. For example, a thin layer of silicon monoxide may be added to protect an aluminized surface.

**Radial distortion**   An alteration in magnification from the center of the field to any point in the field, measured in a radial direction from the center of the field. Some radial distortion is inherent in most optical systems, but can be reduced by proper design.

**Raw glass**   A term that describes any state of glass before its manufacture as an element.

**Reflection**   Return of radiation by a surface, without change in wavelength. The reflection may be specular, from a smooth surface; diffuse, from a rough surface or from within the specimen; or mixed, a combination of the two.

**Reflective coating**   Thin-film coating, single or multilayer, which is applied to surface #2 to increase its reflectance over a specified range of wavelengths. Metal film, transparent or opaque, is the oldest form of reflective coating.

**Reflectivity**   The ratio of the intensity of the total radiation reflected from a surface to the total incident on that surface.

**Refractometer**   An instrument used to measure the refractive index of solids and liquids.

**Roped in**   Term used when the glass is assembled on the body opening by using a rope. The glass in this case will not be bonded to the flange. For roped in assembling, the most common is to have a gasket, but encapsulation can be used too.

**Runout**   In a linear stage, any deviation from the desired translation across a flat, straight line.

**Safety glass**   Safety glazing materials predominantly ceramic in character that meet the appropriate requirements of the safety standard, including (but not limited to) laminated glass and tempered glass.

**Laminated glass**   Two or more pieces of float glass bonded together by an intervening layer or layers of plastic material. It will crack or break under sufficient impact, but the pieces of glass tend to adhere to the plastic. If a hole is produced, the edges are likely to be less jagged than would be the case with ordinary annealed glass.

**Sag**   **1.** In the geometric sense, an abbreviation for the term "sagitta," the height of a curve measured from the chord. It is exact for a parabola. **2.** Conforming a sheet of glass to a ceramic or metal form by heating the glass to its softening point and allowing it to settle.

**Scratch**   A defect on a polished optical surface whose length is many times its width. Block reek is a chainlike scratch formed in polishing. A runner cut is a curved scratch caused by grinding. A sleek is a hairline scratch. A crush or rub is a surface scratch or scratches usually caused by mishandling.

**Scratch-Resistant Coating (SRC)**   Thin layers intended to prevent damage to plastic optics. Used in Poly Carbonates.

**Sidelite**   A side window on a passenger vehicle.

**Spectrophotometer**   An instrument for measuring spectral transmittance or reflectance.

**Surface #1**   In optics, one of the exterior faces of an optical element, e.g. the surface of a windshield that gets wiped, is exposed to the elements

**Surface quality**   The specification of allowable flaws in a surface by comparison to reference standards of quality. Two graded sets of surface quality standards are employed. The first denotes defects of a long nature such as scratches, and the second indicates essentially round defects such as digs.

**Surface quality standards**   The standards of MIL-O-13830 set by the U.S. government relative to tolerable surface scratches and other such defects in an optical system. A series of standard glass plates that have been deliberately scratched with a specific depth and width of scratch can determine the magnitude of the defect in another optical surface by comparison between the two.

**Surface reflection**   Also known as Fresnel reflection. That portion of the incident radiation that is reflected from the surface of a refractive material. It is a function of the refractive index of the material.

**Surface wave**   A wave that is guided by the interface between two different media or by a refractive index gradient in the medium. The field components of the wave may exist (in principle) throughout space, even to infinity, but become negligibly small within a finite distance from the interface. All guided modes, but not radiation modes, in an optical waveguide belong to a class known in electromagnetic theory as surface waves.

**Tangential distortion**   An image defect, usually caused by errors of centration, that results in the displacement of image points perpendicular to a radius from the center of the field.

**Thin film**   A thin layer of a substance deposited on an insulating base in a vacuum by a microelectronic process. Thin films are most commonly used for antireflection, achromatic beamsplitters, color filters, narrow passband filters, semitransparent mirrors, heat control filters, high reflectivity mirrors, polarizers, and reflection filters.

**Tolerancing**   The determination of the degree to which a manufactured component can deviate from its ideal specifications of material and geometry without impairing its performance.

**Transmission (T)**   In optics, the conduction of radiant energy through a medium. Often denotes the percentage of energy passing through an element or system relative to the amount that entered.

**Transmittance**   The ratio of the radiant power transmitted by an object to the incident radiant power.

**Wave**   **1.** An undulation or vibration; a form of movement by which all radiant energy of the electromagnetic spectrum is estimated to travel. **2.** A type of surface defect, usually due to improper processing.

**Wavelength (WL)**   Electromagnetic energy is transmitted in the form of a sinusoidal wave. The wavelength is the physical distance covered by one cycle of this wave; it is inversely proportional to frequency.

**Wedge**   An optical element having plane-inclined surfaces. Usually the faces are inclined toward one another at very small angles. Wedges divert light toward their thicker portions, and may be circular, oblong or square. Wedge PVB (polyvinyl butyral) is what is used to project an image on a windshield for Head-Up Displays (HUDs).

**Wedge tolerance**   A method of specifying the allowable edge-thickness difference or decentering of a lite.

**Window**   **1.** A piece of glass with plane parallel surfaces used to admit light into an optical system and to exclude dirt and moisture. **2.** A particular region of the electromagnetic spectrum that has been singled out for some purpose; for example, the region near 850 nm at which optical fibers operate with lowered losses often is referred to as the first window.

**Wings**   The part of the glass that extends beyound the A-Pillar radius, wrapping around the side of the cap.

# Chapter 14
## Design Rule Tables

The following design tables have been created to provide the reader with a straightforward guide to solve most of the glass-related questions you may face as an engineer. Numeric values as well as ranges denoted by red, yellow, and green indicators have been assigned to let you know what your optimum choice is with its associated risk. Red denotes the highest risk. Yellow indicates medium risk, and green is the lowest risk.

| Table 14.1 Blank Size | | | | |
|---|---|---|---|---|
| **Title** | **Description** | **Design Rule** | **Consequences & Lessons** | **Supporting Information** |
| Windshield Maximum Length in Flat | Measured on Surface 1 (outside surface) must be less than 2387.6 mm (94″). This is the maximum arc length for any high-yield ovens. Arc ≥ 2387.6 mm (94″) will require larger capacity and lower-yield ovens. | Maximum arc length along Surface 1 (outside surface) < 2387.6 mm (94″) (Green)<br><br>2387.6 mm (94″) < Maximum arc length along Surface 1 (outside surface) < x″ (x*25.4 mm/inch) (Yellow) (requires larger oven which can double piece cost)<br><br>Maximum arc length along Surface 1 (outside surface) > x″ (x*25.4 mm/inch) (Red) | Low throughput. Unmet design expectations. A windshield with an arc length greater than 2387.6 mm (94″) will impact tooling and time. The decision to move to a larger, slower oven may impact piece price Δ by > 60%. | Supplier oven capacity |

| Table 14.2 | Forming a Laminated Windshield | | | |
|---|---|---|---|---|
| **Title** | **Description** | **Design Rule** | **Consequences & Lessons** | **Supporting Information** |
| Forming a Laminated Windshield | Measured on Surface 1 (outside surface) must be less than 2387.6 mm (94"). This is the maximum arc length for any high-yield ovens. Arc lengths ≥ 2387.6 mm (94") will require larger capacity and lower yield ovens. Height of the section as measured in Vdoc will also affect supplier choice. | Maximum arc length along Surface 1 (outside surface) < 2387.6 mm (94") (Green)<br><br>2387.6 mm (94") < Maximum arc length along Surface 1 (outside surface) < x" (x*25.4 mm/inch) (Yellow) (requires larger oven which can double piece cost)<br><br>Maximum arc length along Surface 1 (outside surface) > x" (x*25.4 mm/inch) (Red).<br><br>Maximum height ≤ 250 (Green)<br><br>≥ 251 mm to ≤ 414 mm (Yellow)<br><br>≥ 415 mm (Red) | Low throughput. Unmet design expectations. A windshield with an arc length greater than 2387.6 mm (94") will impact tooling and time. The decision to move to a larger, slower oven may impact piece price Δ by > 60%. | Supplier oven capacity |

| Table 14.3 | Forming a Sidelite | | | |
|---|---|---|---|---|
| **Title** | **Description** | **Design Rule** | **Consequences & Lessons** | **Supporting Information** |
| Forming Tempered or Laminated Sidelites | Determine the depth of bend. Is the bend in two axies? Depth of bend will impact cost. | Max Depth of Bend in the Z axis:<br><br>< 10 mm (Green)<br><br>11 mm to 15 mm (Yellow)<br><br>> 15 mm (Red)<br><br>Max Convex in X:<br><br>< 10 mm (Green)<br><br>10 mm to 15 mm (Yellow)<br><br>> 15 mm (Red) | Press Bend—Able to achieve more complex bends<br><br>Sag Bend— Able to contain costs<br><br>Complex curvature requires sufficient packaging envelope in door system. | Capable of producing more complex shapes than sag process |

| Table 14.4 Windshield Shape | | | | |
|---|---|---|---|---|
| Title | Description | Design Rule | Consequences & Lessons | Supporting Information |
| Windshield Surface | Provides optimal viewing and wiping surface | Rate of Change from center line of windshield to EoG, particularly in A-Pillar transition > 600 mm (Green)<br><br>Rate of Change from center line of windshield to EoG, particularly in A-Pillar transition < 600 mm to > 300 mm (Yellow)<br><br>Rate of Change from center line of windshield to EoG, particularly in A-Pillar transition < 300 mm (Red) | Poor Optical Quality<br><br>Wiper Chatter | Design Practice<br><br>Supplier Guidance |

| Table 14.5 Horizontal Depth of Chord | | | | |
|---|---|---|---|---|
| Title | Description | Design Rule | Consequences & Lessons | Supporting Information |
| Windshield—Horizontal Depth of Chord | Horizontal Depth of Chord (Hdoc or Camber) is measured by chording the widest point of windshield after bending. | Hdoc > 381 mm (Red)<br><br>Hdoc 381 mm (Yellow)<br><br>Hdoc < 381 mm (Green) | Poor optics.<br><br>Poor wiper performance.<br><br>Unmet design expectations. A windshield with depth of chord greater than 250 mm will affect supplier selection and piece price cost. | Depth of bend (Chord/Camber) is critical. It can define your supplier choice, yield and cost. Not all glass companies have the same dimensional capacity. The difference in oven height can be as much as 5 in (125 mm). This can affect yield on a normal production run.<br><br>The decision to move to a larger, slower oven may impact piece price $\Delta$ by > 60%. |

| Table 14.6 Vertical Depth of Chord | | | |
|---|---|---|---|
| **Title** | **Description** | **Design Rule** | **Consequences & Lessons** |
| Windshield—Vertical Depth of Chord | Vertical Depth of Chord (Vdoc) is measured by chording the glass at Center Line (CL) | Vdoc > 50 mm (Red)<br>Vdoc 25 mm to 49 mm (Yellow)<br>Vdoc < 10 mm (Green) | Possibly affect optics. Will contribute to poor wiper performance. |

| Table 14.7 Pillar Rise | | | |
|---|---|---|---|
| **Title** | **Description** | **Design Rule** | **Consequences & Lessons** |
| Windshield—Pillar Rise | Pillar rise is measured by chording the glass at edge of glass on the pillar side. | Vdoc > 25 mm (Red)<br>Vdoc < 25 mm but ≥ 19 (Yellow)<br>Vdoc < 19 mm (Green) | Poor wiper performance.<br>Unmet design expectations.<br>Optics. |

| Table 14.8 Pillar Radius | | | |
|---|---|---|---|
| **Title** | **Description** | **Design Rule** | **Consequences & Lessons** |
| Design Rule Pillar Radius | Measured in two locations:<br>1) Approximately 100 mm below the upper edge of glass (EoG) and 100 mm inboard.<br>2) At the point of greatest width on the windshield, usually lower A-Plr, and ~100 mm inboard. | Pillar radius > 250 mm (Green)<br>Pillar Radius 250 mm (Yellow)<br>Pillar Radius < 250 mm (Red) | A radius tighter than 250 mm will cause problems with optics and impact cost. May cause wiper chatter. |

| Table 14.9 Ceramic Frit | | | |
|---|---|---|---|
| **Title** | **Description** | **Design Rule** | **Consequences & Lessons** |
| Frit | Ceramic frit is a paint comprised of minute glass particles and pigment. The enamel is fired onto the glass using a silkscreen process, creating a permanent coating. The application of ceramic frit is often helpful in controlling glare, and providing UV protection. When specified on the engineering print, the glazing shall be supplied with a ceramic coating (frit) applied to the inside surface of the glass to trademark, to provide a decorative appearance, and to prevent light transmittance. | The maximum light transmittance (spectral %) of solid band of ceramic frit shall be 2.0% when measured within the spectrum of 400 to 700 nm per SAE J1976. Use illuminant (daylight), standard observant angle CIE 2° and report the tristimulus value y (Green) | Glass adhesion<br><br>UV degradation<br><br>SAE J1976 |

| Table 14.10 Corner Radii | | | |
|---|---|---|---|
| **Title** | **Description** | **Design Rule** | **Consequences & Lessons** |
| Corner Radii | Styling Cue<br><br>Molding enabler/ inhibitor | Corner radius shall be ≥ 56 mm (Green)<br><br>Corner radius shall be 50 mm to ≤ 55 mm (Yellow)<br><br>Corner radius shall be < 50 mm (Red) | Accommodates many material choices.<br><br>50 mm is the limit to get a molding to bend without puckering.<br><br>Causes puckering in moldings, which may cause leaks. Customer dissatisfier.<br><br>Design Practice.<br><br>Material Choice.<br><br>Supplier Guidance. |

| Table 14.11 Flange Length | | | |
|---|---|---|---|
| **Title** | **Description** | **Design Rule** | **Consequences & Lessons** |
| Minimum Flange Length | Required length of sheet metal flange from tangent to edge.<br><br>Needed to support glass plane and not induce stress riser. | Desired flange length needed to support windshield, backlite, and fixed positions ≥ 22 mm (Green)<br><br>Acceptable flange length needed to support windshield, backlite, and fixed positions ≥ 21 mm (Yellow)<br><br>Unacceptable flange length needed to support windshield, backlite, and fixed positions < 20 mm (Red) | Increased likelihood of part failure |

| Table 14.12 Aspect Ratio | | | |
|---|---|---|---|
| **Title** | **Description** | **Design Rule** | **Consequences & Lessons** |
| Windshield— Aspect Ratio | As the aspect ratio increases, multiple elements of the windshield are affected.<br><br>The greater the depth of chord, the more optics, wiper performance, and tooling are impacted. | AR > .16 (Red)<br><br>AR = .15 (Yellow)<br><br>AR ≤ .10 (Green) | Poor wiper performance. Low throughput furnace, which will result in cost penalty.<br><br>Depth of bend (Chord/Camber) is critical. It can define your supplier choice, yield, and cost. Pilkington has less oven capacity in terms of furnace dimensions (10″ height or 250 mm) than Guardian (15″ or 381 mm).<br><br>The decision to move to a larger, slower oven may impact piece-price Δ by > 60%. |

| Table 14.13 Edge Finish | | | |
|---|---|---|---|
| **Title** | **Description** | **Design Rule** | **Consequences & Lessons** |
| Edge Finish | Edge finish is determined by the position of the window in vehicle (i.e., windshield, door glass, moveable vent, fixed vent with molding) and is rated by quality of finish. The ground edge impacts on the edge stress only for tempered parts. Laminated parts are not affected.<br><br>The reason for grinding is also to remove the sharp edges, and the seamed edge does the same job as the ground.<br><br>Second, the ground edge if compared to seamed, will give you a smoother finishing, and consequently will give you a much better appearance.<br><br>The ground edge is used mainly where the glass is exposed and there are no gaskets to cover. | #1 grind used on exposed edge (Green)<br><br>#2, 3, or 4 grind used on exposed edge (Red)<br><br>Standard Edgework Is SAE # 3. Definition: Semi-crown edge, Semi-satin finish.<br><br>Edge #2 Where the central part of the edge need not be touched with the edging wheel; for edges enclosed in fixed channels, or stationary installations.<br><br>SAE #2 and # 1 available.<br><br>Interior edge radius: minimum interior radius of 200 mm.<br><br>With a minimum lead in radius of 25 mm. | Glass breakage due to edge stress<br><br>Customer dissatisfier<br><br>Personal injury |

| Table 14.14 Edge of Glass to Edge of Hole | | | |
|---|---|---|---|
| **Title** | **Description** | **Design Rule** | **Consequences & Lessons** |
| Edge of Glass (EoG) to Edge of Hole (EoH) | Mitigate stress in a tempered window | Distance shall be $X \geq 1.5 *$ Hole Dia. (Green)<br><br>Distance shall be $X > 1.5 *$ Hole Dia. (Red) | If hole is too close to edge, the part will fail. |

| Table 14.15 Designed Interference—Molding | | | |
|---|---|---|---|
| **Title** | **Description** | **Design Rule** | **Consequences & Lessons** |
| Designed Interference | Design vs. Installed position of windshield molding | Designed interference (installed position) 3 mm to 5 mm (Green)<br><br>Designed interference (installed position) < 3 mm (Red) | Too little interference can cause water leaks and/or wind noise. |

| Table 14.16 Windshield Installation Angle | | | |
|---|---|---|---|
| **Title** | **Description** | **Design Rule** | **Consequences & Lessons** |
| Windshield—Installation Angle | Installation angle is set by Industrial Design. Generally viewed as a styling feature. | Vdoc > 43° (Red)<br><br>Vdoc > 29° to ≤ 42° (Yellow)<br><br>Vdoc ≤ 28° (Green) | It will impact vision, veiling glare, and heat load in the cab. |

| Table 14.17 Windshield Deflection | | | |
|---|---|---|---|
| **Title** | **Description** | **Design Rule** | **Consequences & Lessons** |
| Windshield—Deflection | Glass load<br><br>Tolerancing stack-up | Deflection > 6 mm (Red)<br><br>Deflection > 3 mm to ≤ 6 mm (Yellow)<br><br>Deflection ≤ 3 mm (Green) | Will not meet tolerancing guidelines.<br><br>Potential water leaks.<br><br>Adhesion problems. |

| Table 14.18 UV Transmission | | | |
|---|---|---|---|
| **Title** | **Description** | **Design Rule** | **Consequences & Lessons** |
| UV Transmission | Provides occupant comfort.<br><br>Retards material degradation. | The wavelength range of UV is 300 to 380 nm (Green). | Laminated windshields provide the greatest resistance.<br><br>Long-term UV exposure can result in fading and deterioration of fabrics.<br><br>The only portion of the solar spectrum that is visible to the human eye is the visible light range, 380 to 720 nm. |

| Table 14.19 Light Transmission (AS) requirements | | | |
|---|---|---|---|
| Title | Description | Design Rule | Consequences & Lessons |
| Light Transmission | Acceptable light transmission dictated by window position in vehicle. | AS 1 ≥ 70° (Green)<br>AS 1 ≤ 70° (Red)<br>AS 2 ≥ 70° (Green)<br>AS 2 ≤ 70° (Red) | Illegal for on-road use<br>ANSI Z 26.1-1996 |

| Table 14.20 Infrared Glass | | | |
|---|---|---|---|
| Title | Description | Design Rule | Consequences & Lessons |
| Windshield Design—Infrared-Reflective Glass—Faraday's cage | Infrared-Reflective Glass reduces transmission of UV and IR solar energy, it helps reduce interior heat buildup, shortening cool-down time and reducing heat gain while driving. Consumers do not feel as hot when they enter the vehicle during hot weather, and the vehicle gets to a more comfortable condition quicker once the air conditioner is turned on because it has less heat built up in the vehicle. The main drawback is reflection of signals by a transponder on some interstate toll ways. | This is a real concern when cars and trucks utilize IRR glass. The IRR technology does an effective job of reflecting infrared energy so the occupants stay cooler, but the reflective surface is metalized and can electro-magnetically shield the interior.<br>As a point of reference: if there is ever interest in IRR technology in the future, there are ways around this shielding effect that can be designed into the windshield.<br>It's not heated glass that causes the issue, it's the IRR coating.<br>There is a hole cut into the IRR surface by design that makes a leak path to the Faraday cage into the windshield. | Glass adhesion<br>UV degradation |

| Table 14.21 Heated Glass | | | |
|---|---|---|---|
| Title | Description | Design Rule | Consequences & Lessons |
| Heated Glass Systems | Provide adequate defog/defrost capabilities to all glazing components. | Buss bars & Grid lines ≥ 20 mm from bonding path (Green)<br>Buss bars and Grid lines ≥ 15 mm to < 20 mm from bonding path (Yellow)<br>Buss bars and Gridlines < 15 mm from bonding path (Red) | Bond failure |

| Table 14.22 Urethane Bead Height | | | |
|---|---|---|---|
| **Title** | **Description** | **Design Rule** | **Consequences & Lessons** |
| Urethane Bead Height | The windshield urethane thickness (height) or bond gap target should be 14 to 18 mm.<br><br>The windshield urethane width target should be 12 to 15 mm.<br><br>The bead will appear on the drawing as an inverted diamond. | Bead height 15 mm (Green)<br><br>Bead height >10 and < 15 mm (Yellow)<br><br>Bead height < 10 mm (Red) | Windshield breakage<br><br>Flange read-out<br><br>Water leaks<br><br>Loss of adhesion |

| Table 14.23 Moldings—Fixed Glass | | | |
|---|---|---|---|
| **Title** | **Description** | **Design Rule** | **Consequences & Lessons** |
| Moldings—Fixed Glass | The fixed glass seal may press or be bonded onto the edge of glass, and door inner. The seal/molding may have molded corner transitions.<br><br>The seal will prevent air and water leaks when the window is closed throughout all tolerance variations. | Designed interference (tail on molding) to body ≥ 3 mm (Green).<br><br>Designed interference (tail on molding) to body ≤ 3 mm (Yellow).<br><br>Designed interference (tail on molding) to body < 2 mm (Red).<br><br>Gap to Perpendicular Flange ≤ 5 mm (Green).<br><br>Gap to Perpendicular Flange ≥ 6 mm (Yellow).<br><br>Gap to Perpendicular Flange > 7 mm (Red).<br><br>Gap to Parallel (to glass surface) Flange 6 mm (Green).<br><br>Gap to Parallel (to glass surface) Flange > 6 mm < 8 mm (Yellow).<br><br>Gap to Parallel (to glass surface) Flange ≥ 8 mm (Red). | Will not meet tolerancing guidelines.<br><br>Potential water leaks.<br><br>Adhesion problems. |

**Table 14.24  Molding Criteria**

| Title | Description | Applies to | Design Rule | Consequences & Lessons |
|---|---|---|---|---|
| Molding Criteria | Material choice based on geometry and performance criteria | This design practice applies to all commercial truck, windshield applications. | Slow corner ≥ 56 mm/softer durometer (Green)<br><br>Accelerated corner ~51 mm to 53 mm/hard material (Red) | Soft materials run the risk of popping off the glass.<br><br>Caution is needed. Accelerated corner, hard material will cause puckering.<br><br>Muckets may be needed, which is visually unappealing. |

**Table 14.25  Seals—Vent Window**

| Title | Description | Design Rule | Consequences & Lessons |
|---|---|---|---|
| Seals—Vent Window | The vent window seal will press onto the weld flange created by the door outer and door inner. The seal may have molded corner transitions. The seal will prevent air and water leaks when the window is latched closed throughout all tolerance variations. | Gap Range > 16.5 mm (Red)<br><br>Gap Range 15 mm to 16.5 mm (Yellow)<br><br>Gap Range < 15 mm (Green) | Will not meet tolerancing guidelines.<br><br>Potential water leaks.<br><br>Adhesion problems. |

**Table 14.26  Gaskets—Windshield**

| Title | Description | Design Rule | Consequences & Lessons |
|---|---|---|---|
| Gaskets—Windshield | Provides water tight and debris free seal between windshield and cab opening. Meets Industrial Design criteria. Provides smooth transition between straight line elements and corners. | Corner Radius < 50 mm (Red)<br><br>Corner Radius 50 mm (Yellow)<br><br>Corner Radius > 50 mm (Green) | Puckering in corners if radius is too tight.<br><br>Potential for gapping and water leaks. |

| Table 14.27 | | Glass Support—Windshield | |
| --- | --- | --- | --- |
| Title | Description | Design Rule | Consequences & Lessons |
| Support—Glass Systems | To provide adequate support to all glazing for specified time, allowing adhesive to set. May also provide locating feature. | Windshield support made from a polymer (material softer than glass). Contact edge must be straight or parallel to glass edge (Green). Contact edge not straight or parallel to glass edge (Yellow). Contact edge makes sharp contact with edge of glass (Red). | Stress must be distributed along glass edge as uniformly as possible. If the contact point is narrow, a stress riser will be created. Glass is most susceptible to failure within the first 20 mm from its edge. |

| Table 14.28 | | Glass Support Fixed Position | |
| --- | --- | --- | --- |
| Title | Description | Design Rule | Consequences & Lessons |
| Support—Glass Systems | To provide adequate support to all glazing components for specified time to allow adhesive to set. May also provide locating feature. | Fixed glass support maybe made from a polymer (material softer than glass). Must provide support or locating fixture (Green). Interferes with bonding. Increases gap which may cause water leaks. Support produces a stress riser (Red). | Stress should be distributed uniformly along the glass edge. If the contact point is narrow, a stress riser will be created. If stops are of the pin type, they must be secured to the surface (generally fixed vent) and fit securely in receiving hole (sheet metal). |

| Table 14.29 | | Wiper Quality—Interface Control Document | |
| --- | --- | --- | --- |
| Title | Description | Design Rule | Consequences & Lessons |
| Wipe Quality | Provides optimal viewing and wiping surface | Design Target Arm Load = ~15 N/m; = ~13 N/m for beams (Green) Design Target Arm Load > 13 N/m for beams (Yellow) It is hard to say when unmet design expectations will occur. Load can be increased to minimize lift. | Flatter glass improves wiper performance. The guidance for arm load is for typical wiper system and, if possible, should be maintained throughout the wipe arc under various wind load conditions. The effect of deviation from nominal depends on the system and blade design (pressure distribution). Poor Optical Quality. Wiper Chatter. |

| Table 14.30 Wiper Quality—Interface | | | |
|---|---|---|---|
| **Title** | **Description** | **Design Rule** | **Consequences & Lessons** |
| Wiper Key Points | Provides optimal viewing and wiping surface | Wiper linkages straight arm type (Green)<br><br>Wiper linkages bent arm type (Yellow)<br><br>Wiper linkages multiple bend arm type (Red) | Increased system costs drive greater tolerances.<br><br>Wiper Chatter.<br><br>Motor and levers should be planar. A motor out of plane driving a connecting arm at an angle will result in an elliptical drive motion that will drive up system loads and throw linkage accelerations out. |

| Table 14.31 Wiper Zones Single Pane | | | |
|---|---|---|---|
| **Title** | **Description** | **Design Rule** | **Consequences & Lessons** |
| Meeting ABC wipe zones based on SAE J198 | Meeting ABC wipe zones on a single pane windshield | Zone A $\geq$ 90% (Green)<br>Zone A < 90% (Red)<br>Zone B $\geq$ 96% (Green)<br>Zone B < 96% (Red)<br>Zone C = 100% (Green)<br>Zone C < 100% (Red) | Failure to meet this requirement may result in failure to offer saleable vehicle.<br><br>SAE J198 |

| Table 14.32 Wiper Zones Multiple Panes | | | |
|---|---|---|---|
| **Title** | **Description** | **Design Rule** | **Consequences & Lessons** |
| Meeting ABC wipe zones based on SAE J198 | Meeting ABC wipe zones on multiple pane windshield | Zone A $\geq$ 65% (Green)<br>Zone A < 65% (Red)<br>Zone B $\geq$ 70% (Green)<br>Zone B < 70% (Red)<br>Zone C = 84% (Green)<br>Zone C < 84% (Red) | Failure to meet this requirement may result in failure to offer saleable vehicle. |

| Table 14.33  Joint Movement | | | |
|---|---|---|---|
| **Title** | **Description** | **Design Rule** | **Consequences & Lessons** |
| Joint Movement | Movement to joint width guideline:<br><br>10% movement materials—butyl rubbers and acrylic latex sealants.<br><br>Movement × 10 = joint width 25% movement materials—polyurethanes and silicones.<br><br>Movement × 4 = joint width 50% movement materials—polyurethanes and silicones. | Butyl ≤ 10% (Green)<br><br>Rubbers & Acrylics ≤ 25% (Green)<br><br>Polyurethanes ≤ 50% (Green) | Failure to meet this requirement may result in failure to seal out unwanted moisture or bond failure. |

| Table 14.34  Encapsulated Glass Capabilities | | | |
|---|---|---|---|
| **Title** | **Description** | **Design Rule** | **Consequences & Lessons** |
| Encapsulated Glass Capabilities | Polyurethane material per part can vary from a minimum of 40 grams to a maximum of 1500 grams.<br><br>This is dependent on part size, design, and geometry. | Width: 8″ to 80″ (204 mm to 2050 mm)<br><br>Height: 8″ to 35″ (204 mm to 888 mm)<br><br>Max Depth of Bend: 9″ (228 mm) | Failure to meet this requirement may result in failure to seal out unwanted moisture or bond failure. |

| Table 14.35  Thermal Shock | | | |
|---|---|---|---|
| **Title** | **Description** | **Design Rule** | **Consequences & Lessons** |
| Thermal Shock | Glass will break by thermal shock, not by thermal conditions alone. | Properly ground glass edge<br><br>Controlled rate of temperature increase | Failure to meet these requirements may result in failure of component in the field. |

**Table 14.36    Optical Considerations**

| Title | Description | Design Rule | Consequences & Lessons |
|---|---|---|---|
| Curved Glass and Optical Considerations | Accelerated curvature will cause distortion. | No curvature tighter than 50 mm acceptable.<br><br>Minimize rate of change. | As the curve increases the distortion increases, because an object ceases to be viewed normal (90°). The viewer now sees an object through the thickness of the glass, which increases with curvature. |

**Table 14.37    Glass Curvature**

| Title | Description | Design Rule | Consequences & Lessons |
|---|---|---|---|
| Curved Glass and Optical Considerations | Accelerated curvature will cause distortion. | No curvature tighter than 50 mm acceptable.<br><br>Minimize rate of change. | Poor optics, improper molding fit |

**Table 14.38    Glass Edge Grind #1**

| Title | Description | Design Rule | Consequences & Lessons |
|---|---|---|---|
| #1 edge grind | Edge finish for any glass that may be touched by hand | Edgework # 1.<br><br>Round satin finish for exposed installation.<br><br>No shell chips.<br><br>No shiners.<br><br>No sharp edges.<br><br>No edge peel.<br><br>Grind must be centered. | Improperly ground edge results in customer dissatisfaction. |

| Table 14.39 Glass Edge Grind #2 | | | |
|---|---|---|---|
| **Title** | **Description** | **Design Rule** | **Consequences & Lessons** |
| #2 edge grind | Edge finish for any glass that will be covered by a molding | Edgework # 2.<br><br>Round semi-satin finish for unexposed installation.<br><br>3 mm shell chip, 1 per 150 mm. Max 4 per side.<br><br>Shiners: ⅓ edge width, 50 mm in length, 1 per 150 mm.<br><br>No sharp edges.<br><br>#2 edge is created with a diamond impregnated grinding wheel. This is the same as #1 edge, just run on the line at a faster rate. | Use where hidden from view or touch |

| Table 14.40 Glass Edge Grind #3 | | | |
|---|---|---|---|
| **Title** | **Description** | **Design Rule** | **Consequences & Lessons** |
| #3 edge grind | Edge finish for any glass that will be covered by a molding | Edgework # 3<br><br>Semi-satin finish, usually for stationary installation.<br><br>6 mm shell chip, 1 per 150 mm, maximum 4 per side.<br><br>Shiners: up to ¾ edge width.<br><br>Edge peel 3 mm × 250 mm.<br><br>No sharp edges. | Use where hidden from view or touch.<br><br>Approximate cost savings of $3.00 per part over finer grinds. |

| Table 14.41 Glass Edge Grind #4 | | | |
|---|---|---|---|
| Title | Description | Design Rule | Consequences & Lessons |
| #4 edge grind | Edge finish for any glass that will be covered by a molding, encapsulation | Edgework # 4. Seamed edge finish for unexposed installation. 6 mm shell chips, 1 per 150 mm, max 4 per side. 8 mm min to 1.5 mm max seam width. No sharp edges. | Absolutely no exposed edge. Used primarily in windshield applications. Least expensive finish. |

| Table 14.42 Heated Windscreens | | | |
|---|---|---|---|
| Title | Description | Design Rule | Consequences & Lessons |
| Heated Windscreen | Wired parts must meet regulatory standard (e.g., ECE) requirements (as for non-heated laminate)—Bake Test, Boil Test, head form, plus any additional durability tests. | Wired Heated Windscreen: Product validation and testing product design (Defrost). Parts must meet legal ECE 78/313 directive for defrost/de-mist requirements. Directive applicable to any combination of heater/blower/wiper system. Test conditions—iced windscreen (0.044 g/cm$^2$ water) at −0.4°F (−18°C). Requires 80% of vision area "A" defrosted after 20 min (use of wiper blades permitted). Test carried out on fully glazed vehicle. To meet requirements and OEM expectations—typical design of 500-600 W/m$^2$. Heated screen only—no blowers/wipers used. | Will not meet Design Requirements (DRD) set forth by OEM. May deliver too much electrical energy that produces thermal shock. May not deliver enough heat to defrost/defog a given area over time. |

| Table 14.43 Product Validation for Wired Parts | | | |
|---|---|---|---|
| Title | Description | Design Rule | Consequences & Lessons |
| Validating Wired Parts | Wired parts must meet regulatory standard (e.g., ECE) requirements (as for non-heated laminate)—Bake Test, Boil Test, Head Form, plus additional durability tests. | 50 day 5% salt mist exposure test (conditions as DIN50021) with cyclic powering 30 min on/30 min off.<br><br>Multi-stage high temperature/humidity exposure with cyclic powering (20 day duration) with temperature exposure –22°F to +122°F (–30°C to +50°C) (with 97% humidity).<br><br>No change in resistance >7.0% or change in number of inoperable wires allowed after testing.<br><br>Connectors must meet typical 60-N pull off force without detachment from windscreen. | Will not meet Design Requirements (DRD) set forth by OEM |

| Table 14.44 Antennas | | | |
|---|---|---|---|
| Title | Description | Design Rule | Consequences & Lessons |
| Antennas | Mast | Increased wind noise.<br><br>Better warranty.<br><br>Less appealing.<br><br>Lower costs. | Poor wiper performance.<br><br>Unmet design expectations.<br><br>Optics. |
| | Integrated | Improved appearance.<br><br>Better wind tunnel performance.<br><br>Higher costs. | Printing the silver antenna may result in an additional cost of up to $20 per vehicle; there could also be amplifiers necessary and potentially design costs. |

| Table 14.45    AS Designations | | | |
|---|---|---|---|
| **Title** | **Description** | **Design Rule** | **Consequences & Lessons** |
| AS Designations | AS1 | Safety glazing materials that must be used anywhere in a motor vehicle and must have a visible light transmittance of at least 70°. | These are federal mandated specifications. |
| | AS2 | Safety glazing material for use anywhere in a motor vehicle except windshields. This product cannot have a visible light transmittance of less than 70°. | |
| | AS3 | Safety glazing material for use anywhere in a motor vehicle except windshields and certain specified locations that are requisite for driving visibility such as the immediate right and left of the driver. Typically this product has less than 70° light transmittance, but does not have to be, again dependent on vehicle position. | |

| Table 14.46    Glass Tolerance | | | |
|---|---|---|---|
| **Title** | **Description** | **Design Rule** | **Consequences & Lessons** |
| Tolerances for Curved and Flat Glass | Windshields (Curved Glass) | Size: tolerance range 3 mm for curved and 2 mm for flat glass. Size: when glass is curved a greater tolerance is necessary due to the bending and tempering process. In flat glass the size change is minimal after tempering.<br><br>Form: a form tolerance is required for flat glass due to warping after the tempering process.<br><br>Thickness tolerance:<br><br>For laminated glass the tolerance is ±0.4 mm due to stack up tolerances for each of the glass sheets.<br><br>Cross curvature tolerance:<br><br>This is the point measured at the center of the glass: ±4 mm.<br><br>Ceramic frit tolerance:<br><br>Position: ±1.5 mm<br><br>Frit dimensions: ±1 mm<br><br>Form: tolerance range 3 mm for curved and 1.5 mm for flat glass. Size: tolerance range is 2 mm for curved and 1.5 mm for flat glass. | Water leaks into cabin, noise. |
| | Sidelites and backlites (tempered) | Holes:<br><br>Position: ±1 mm<br><br>Diameter: ±0.5 mm | |

| Table 14.47 Regulatory Requirements | | | |
|---|---|---|---|
| **Title** | **Description** | **Design Rule** | **Consequences & Lessons** |
| Requirements for Glass and Wipers | FMVSS 205<br><br>CMVSS 205<br><br>FMVSS 104<br><br>CMVSS 104<br><br>ECE R43<br><br>SAE J198 | Products intended to be produced for North America should be designed to meet the applicable regulatory requirements FMVSS 205 (glazing material), CMVSS 205, FMVSS 104 (windshield washing and wiping), CMVSS 104, and to also meet IAC-003 (international acceptance criteria for FMVSS/CMVSS 104, which we treat as a regulatory requirement); as well as all applicable state regulations.<br><br>Products intended also be exported outside of North America should also be designed to the guidelines mentioned below.<br><br>The other countries around the world require vehicle glass to comply with ECE 43.<br><br>There's no ECE requirement for wash/wipe for trucks or buses. | Fail to meet, fail to sell |

| Table 14.48 Technical Properties of Glass | | | |
|---|---|---|---|
| **Title** | **Description** | **Design Rule** | **Consequences & Lessons** |
| Technical Properties of Glass | Modulus of Elasticity (Young's) | 10.4 × 106 psi (7.2 × 10 Pa) | |
| | Modulus of Rigidity (Shear) | 4.3 × 106 psi (3.0 × 10 Pa) | |
| | Bulk Modulus | 6.2 × 106 psi (4.3 × 10 Pa) | |
| | Poisson's Ratio | 0.23 | |
| | Density | 158 lb/ft$^3$ (0.00253 g/mm$^3$) | |
| | Coefficient of Thermal Stress | 50 psi/°F (0.62 MPa/°C) | |
| | Thermal Conductivity at 75°F (23.89°C) | 6.5 (Btu)(in)/(h)(°F)(ft$^2$) (0.937 W/m°C) | |
| | Specific Heat at 75°F (23.89°C) | 0.21 Btu/lb$_m$°F (0.88 kJ/kg°C) | |
| | Coefficient of Linear Expansion (75 to 575°F (23.89 to 301.67°C)) | 4.6 × 10$^{-6}$in/in°F (8.3 × 10$^{-6}$mm/mm°C) | |

**Table 14.49 Technical Properties of Polyvinyl Butyral**

| Title | Description | Design Rule | Consequences & Lessons |
|---|---|---|---|
| Technical Properties of Polyvinyl Butyral | Modulus of Rigidity (Shear) | .6 to 4.4 MPa | |
| | Modulus of Elasticity (Young's) | 2 to 13 MPa | |
| | Poisson's Ratio | 0.5 (@ 73.4°F (23°C)) | |
| | Density | 1.04 g/cm$^3$ | |
| | Coefficient of Linear Expansion | 560 × 10$^{-6}$ mm/mm°C | |
| | Specific Heat | 1980 J/kg K | |
| | Thermal Conductivity | 0.2 W/mK | |

**Table 14.50 Tempered Glass Criteria**

| Title | Description | Design Rule | Consequences & Lessons |
|---|---|---|---|
| Tempered Glass Criteria | Federal safety laws require that window glass be tempered if each of the following criteria are met. | Sill height within 18 in (0.457 m) of the floor, top edge greater than 36 in (0.914 m) from the floor, area greater than 9 ft$^2$ (0.836 m$^2$), and horizontal distance to nearest walking surface of less than 36 in (0.914 m). | Failure to pass federal safety standards, part must be replaced and repaired to meet test. |

**Table 14.51 Extrusions**

| Title | Description | Design Rule | Consequences & Lessons |
|---|---|---|---|
| Extrusions | EPDM | Advantages:<br>  Less compression set<br><br>Disadvantages:<br>  Longer cycle time<br>  Higher tooling cost<br>  Not recyclable<br>  Poor color control | Material leader in glass run channels |
| | TPV | Advantages:<br>  Shorter cycle times<br>  Less overall cost<br>  OEM approved<br><br>Disadvantages:<br>  More compression set | |

| Table 14.52   Seals | | | |
|---|---|---|---|
| Title | Description | Design Rule | Consequences & Lessons |
| Seals | EPDM | Advantages:<br>   Less compression set<br><br>Disadvantages:<br>   Longer cycle time<br>   Higher tooling cost<br>   Not recyclable<br>   Poor color control | Minimum force of 2 N/100 mm. A seal after extrusion starts to age and the stress-strain curve changes. Over a period of time the modulus deteriorates, causing the seal to exert less force on the window. A seal under constant load develops a permanent set below 2 N/100.<br><br>No molded corners present possible leaks.<br><br>75 durometer base material makes for difficult installation.<br><br>Insufficient slip coat thickness may cause seal rollover. |
| | TPV | Advantages:<br>   Shorter cycle times<br>   Less overall cost<br>   OEM approved<br><br>Disadvantages:<br>   More compression set | |

| Table 14.53   Primer Path | | | |
|---|---|---|---|
| Title | Description | Design Rule | Consequences & Lessons |
| Primer Path | Clear Primer | 25 mm ±2 (Typ) based off of urethane bead $C_L$ | Failure to apply or not to apply enough will cause glass to fail from improper bonding/adhesion. |
| | Black Primer | 20 mm ±2 (Typ) based off of urethane bead $C_L$ | Black primer is need for UV protection and abating molding deterioration. Also provides an aesthetic feature. |

| Table 14.54   Laminated Sidelites | | | |
|---|---|---|---|
| Title | Description | Design Rule | Consequences & Lessons |
| Laminated Sidelites | Inner lite | 1.6 mm to 2.1 mm | Improves security; i.e., smash and grabs.<br>Improves safety by limiting flying glass if the window fails in the field.<br>Reduces wind noise by up to 6 dB.<br>Enables increased styling options.<br>Eliminates 98% of UV rays (SPF 50+).<br>IRR enabler. |
| | PVB | 0.6 mm to .76 mm | |
| | Outer lite | 1.6 mm to 2.1 mm | |

| Table 14.55 Photovoltaic Glass | | | |
|---|---|---|---|
| **Title** | **Description** | **Design Rule** | **Consequences & Lessons** |
| Photovoltaic Glass | Reduces heat load into cabin | Limited use in passenger vehicles | Approximately 10 microseconds, virtually instantaneous. |
| | | | PU RIM encapsulation possible. |
| | | | Operating range, −86 to +230°F (−30 to +110°C), LC durability enhanced by UV absorbing front protective layer. |
| | | | Sound attenuation superior to acoustic laminates. |
| | | | Governed by base construction and unchanged by switching state. Wide range of performance possible, solar reflective coatings optional. UV Trans near zero. |
| | | | Governed by construction, 10% to 75% LTa available. Limitless construction variants. |
| | | | Reduces heat load in cabin. |
| | | | Primarily used in sunroofs on RVs. |
| | | | Unusable for passenger car, truck, or bus application. |
| | | | ~10% return on investment. |

| Table 14.56 Seal Construction—Horizontal Seals | | | |
|---|---|---|---|
| **Title** | **Description** | **Design Rule** | **Consequences & Lessons** |
| Seal Construction —Horizontal Seals | Design interference between seal and sliding glass is 3.0 mm. | Design intent at nominal condition is to produce Compression, Load, and Deflection (CLD) forces at 5 N to 7 N per 100 mm. | Failure in field including water intrusion, and squeaks and rattles from sheet metal. |
| | | Minimum CLD to be 2 N per 100 mm at worst case gap. | |
| | | Seal profile needs to be indicated on drawing as a critical characteristic and must be controlled by the supplier. | |

| Table 14.57 Seal Construction—Y Seals | | | |
|---|---|---|---|
| **Title** | **Description** | **Design Rule** | **Consequences & Lessons** |
| Seal Construction—Y Seals | Design interference between seal and sliding glass is 3.0 mm. | Design intent at nominal condition is to produce Compression, Load, and Deflection (CLD) forces at 5 N to 7 N per 100 mm.<br><br>Minimum CLD to be 2 N per 100 mm at worst case gap.<br><br>Seal profile needs to be indicated on drawing as a critical characteristic and must be controlled by the supplier. | Failure in field including water intrusion, and squeaks and rattles from sheet metal. |

| Table 14.58 Seal Construction—Coating Areas | | | |
|---|---|---|---|
| **Title** | **Description** | **Design Rule** | **Consequences & Lessons** |
| Seal Construction—Coating Areas | Slip Coat. (Slip Coat is the generic name of a series of functional water-based coatings used to coat a variety of extruded materials.) | Preferred method<br><br>Coating thickness: Co-Extruded = 80 Microns (Min)<br><br>Coating thickness needs to be indicated on drawing as critical characteristic and must be controlled by the supplier. | Can be done on line.<br><br>Passes OEM lifecycle requirement.<br><br>No capital expense; the coating is co-extruded onto the seal. |
| | Flocking. (Flocking can be used for insulation to foreign elements.) | | Can be done on line.<br><br>Messy procedure.<br><br>Large capital expense if the line is not already set up for it.<br><br>Passes OEM lifecycle requirements. |

| Table 14.59 Locating Pins | | | |
|---|---|---|---|
| **Title** | **Description** | **Design Rule** | **Consequences & Lessons** |
| Locating Pin | Secures the glass to the sheet metal. | Tolerancing may be tight.<br><br>Mostly made from a Nylon derivative.<br><br>Attached to the glass with an adhesive.<br><br>Must be able to hold at least 100 lb. | A realistic work life of approximately 20 minutes. It is needed primarily to hold the glass in place while the urethane sets up. |

| Table 14.60 Datums | | | |
|---|---|---|---|
| **Title** | **Description** | **Design Rule** | **Consequences & Lessons** |
| Windshield | Windshields (laminated) | Best fit: used when the glass will be assembled with gaskets or moldings. There are no specific points of contact between the glass and the opening. The best fit is the method used with a noncomplex design.<br><br>The dimension 24.264" in the example can vary ±0.06 inches. | Must be to print. |

| Table 14.61 Field Failure—Delamination | | | |
|---|---|---|---|
| **Title** | **Description** | **Design Rule** | **Consequences & Lessons** |
| Field Failure—Delamination | Windshields (laminated) | Monitor all phases of production from the autoclave process to molding type and fit. Encapsulation.<br><br>Heating elements. | Finger delamination looks like a hand or fingers that cause pockets of air/moisture in the polyvinyl butyral (PVB). A possible cause of moisture in the inner layer is environmental moisture that becomes trapped between the glass/encapsulation and the sheet metal opening. |

| Table 14.62 Field Failure—Fogging | | | |
|---|---|---|---|
| **Title** | **Description** | **Design Rule** | **Consequences & Lessons** |
| Field Failure—Fogging | Fogging between panes. | Glass must be properly sealed in frame. Desiccant may be used between panes.<br><br>Must survive humid conditions for extended periods of time. | Field Example: Side entrance door glass was fogging up due to poor seal and extremely humid conditions. |

| Table 14.63 Material Storage | | | |
|---|---|---|---|
| Title | Description | Design Rule | Consequences & Lessons |
| Material Storage | Proper storage | Quality control when it comes to material storage and handling is critical to a robust product. A thermal blanket can be used to keep the urethane at the proper temperature, ~85°F. | Improper storage will lead to poor adhesion. Urethane and primer must use first-in, first-out methodology. Material must be stored indoors at ~85°F. |

| Table 14.64 Windshield Forming Capability | | | |
|---|---|---|---|
| Title | Description | Design Rule | Consequences & Lessons |
| Windshield Forming | Sag | ≤ 3048 mm—Width<br>≤ 1880 mm—Height<br>≤ 558 mm—Depth of Bend<br>≥ 250 mm—Radius at A-Pillar | Sag is lower cost, good choice for less complex shapes. |
| | Press | ≤ 2082 mm—Width<br>≤ 1168 mm—Height<br>≤ 35 mm—Convex<br>≤ 200 mm—Radius at A-Pillar | |

| Table 14.65 Polycarbonate Glazing | | | |
|---|---|---|---|
| Title | Description | Design Rule | Consequences & Lessons |
| Polycarbonate Glazing | Lightweight glazing material for specific applications. | Should only be used on a fixed vent application on an automobile or any other commercial vehicle. | It is anywhere from 40-60% lighter in weight than glass, resulting in improved fuel economy.<br>It can be injection molded into complex shapes and designs.<br>It can be produced with distinctive colors and textures.<br>It permits parts integration.<br>It is very tough, protecting vehicle occupants in accidents.<br>It offers improved security against theft since it does not shatter as easily as glass.<br>May offer a price penalty of three times higher than the original price.<br>Limited in use on vehicle. |

| Table 14.66 Banded Glass | | | |
|---|---|---|---|
| **Title** | **Description** | **Design Rule** | **Consequences & Lessons** |
| Banded edge of glass | Raw edge must be covered with a molding of a given type. No specific type exists. During the timeframe that this was written, banding referred to a U channel made out of metal. | No exposed edge of glass on any position for a vehicle to be used to transport children; i.e., school bus. | OEM will not be allowed to sell in that state. This information is found in SAE ANSI Z26.1. |

| Table 14.67 Windshield Installation Tooling | | | |
|---|---|---|---|
| **Title** | **Description** | **Design Rule** | **Consequences & Lessons** |
| Windshield Installation Tooling | Hydraulically applied tooling. | Uniform force in a consistent vector around the periphery is needed to properly install the windshield into the body opening. | Pneumatically controlled installation assists are part of the installation tool that places the windshield on the body opening. These tools are more expensive so you will find them in more automotive plants than anywhere else. This is due to a higher production run which enables the OEM to allocate more capital expenditures to the factory floor. The greatest advantage to this is equal distribution of force, which positions the windshield more uniformly in the body opening. |

| Table 14.68   Manually Applied Windshield Installation | | | |
|---|---|---|---|
| **Title** | **Description** | **Design Rule** | **Consequences & Lessons** |
| Windshield Installation Tooling | Manually applied installation. | Uniform force is needed to properly install the windshield into the body opening. | The risk of a manually applied install is the unequal distribution of force. This may be aided by a two-person install where the installers are of different heights. The force applied is the same but the vector is not.<br><br>Used mainly in low production operations where capital tooling expenditures are more sensitive. |

| Table 14.69   Windshield Racking | | | |
|---|---|---|---|
| **Title** | **Description** | **Design Rule** | **Consequences & Lessons** |
| Windshield Racking | Windshield storage before shipping. | Foam spacers must be placed between stacked windshields. The glass must also rest on a padded surface. | Product failure before shipping. Chipped edges that may cause chipping or cracking in vehicle. |

| Table 14.70   Seal Deflection—Door Systems | | | |
|---|---|---|---|
| **Title** | **Description** | **Design Rule** | **Consequences & Lessons** |
| Seal deflection— Door Systems | Includes requirements for separate front door and rear door above belt glass run seals.<br><br>The purpose of the glass run seal is to retain and guide the drop glass in all conditions from full up to full down. The glass run seal will also seal the cab from water and air intrusion, contribute to inside cab noise reduction, and provide a low friction surface on which the glass can move up or down. | Deflection > 6 mm (Red)<br><br>Deflection < 6 mm (Yellow)<br><br>Deflection 3 mm (Green) | Will not meet tolerancing guidelines.<br><br>Potential water leaks.<br><br>Adhesion problems. |

| Table 14.71 Lamination Inspection Guidelines—Antenna | | | |
|---|---|---|---|
| **Title** | **Description** | **Design Rule** | **Consequences & Lessons** |
| Antenna | Antenna air | Any air along the antenna wire<br>1.5 mm × 5.0 mm—Maximum affected area. Must fit within a 1.5 mm × 5.0 mm grid. | Will not meet customer expectations.<br>All rules pertain to "A" zone.<br>Refer to SAE J198 for zones. |
| | Antenna wire orientation | 15.0 mm Maximum variance allowed measured from the top center of the cutout to the opposite center of the windshield. | |
| | Burn marks—areas of discoloration in the vinyl caused by the antenna iron | 5.0 mm Maximum deviation for antenna wire—not to interfere with vision. | |

| Table 14.72 Lamination Inspection Guidelines—Glass Form | | | |
|---|---|---|---|
| **Title** | **Description** | **Design Rule** | **Consequences & Lessons** |
| Glass Form | Convex | The sag or curve of the windshield.<br>Refer to part specification. | Will not meet customer expectations.<br>All rules pertain to "A" zone.<br>Refer to SAE J198 for zones |
| | Difference between lites (bent only) | Difference in size between the small and large lites.<br>Refer to part specification. | |
| | Glass size to tin (flat only) | Refer to part specification. | |
| | Kink/rate of change | Measure and classify the deviation of a bent piece of glass on a check fixture over a given distance—measured as rise over run. | |
| | Off form | Allowance for an over bent or under bent condition on the check plastic. | |
| | Reverse | Refer to part specification. | |
| | Wavy cut | Any abrupt change, positive or negative, in the accepted curvature of flatness of the glass.<br>2.3 mm maximum.<br>Cannot exceed 100 mm in length. | |

| Table 14.73 Lamination Inspection Guidelines—Edge Quality | | | |
|---|---|---|---|
| **Title** | **Description** | **Design Rule** | **Consequences & Lessons** |
| Edge Quality | Edge Work | A seamed edge so that no chipped edges or fractures may occur.<br><br>No sharpness allowed (unless otherwise specified). | Will not meet customer expectations.<br><br>All rules pertain to "A" zone.<br><br>Refer to SAE J 198 for zones. |
| | Edge Chip | The flaking off of small particles of glass, usually on the edge, which impairs glass strength.<br><br>5 mm circle allowed.<br><br>Depth: Not to exceed ½ thickness of the affected lite.<br><br>Can rework up to 6 mm chip for OEM.<br><br>Rework 7 mm chip for aftermarket. | |
| | Edge Peel | Maximum of 3 mm × 100 mm in length. Total accumulated length: 250 mm.<br><br>Depth: Not to exceed ¼ thickness of the affected lite. | |
| | Fish Hook | Improper scoring or cutting, parts have a curved score mark from the cutting process.<br><br>None allowed. | |

| Table 14.74 Lamination Inspection Guidelines—Inside Contamination | | | |
|---|---|---|---|
| **Title** | **Description** | **Design Rule** | **Consequences & Lessons** |
| Inside Contamination | Cosmetic imperfections<br>Foreign objects<br>Black spots<br>Digs<br>Surface tin<br>Rubs<br>Paint in body<br>Adhesion chips<br>Seeds, metal | Must be contained within a 1 mm circle or up to 2 mm in length if less than 0.5 mm wide (Green). | Will not meet customer expectations.<br>All rules pertain to "A" zone.<br>Refer to SAE J 198 for zones. |
| | Lint, Thread, Hair | Maximum of 10.0 mm in length. | |
| | Stain | Light: Not visible to standard inspection.<br>Allowed not to interfere with vision. | |

| Table 14.75 Lamination Inspection Guidelines—Other Visual Defects | | | |
|---|---|---|---|
| **Title** | **Description** | **Design Rule** | **Consequences & Lessons** |
| Inside Contamination | Air behind mirror mount. | 3.0 mm maximum; must be enclosed on all sides; no more than one defect allowed. | Will not meet customer expectations.<br>All rules pertain to "A" zone.<br>Refer to SAE J 198 for zones. |
| | Air bubbles (bubbles in plastic inner layer or between plastic inner layer and glass). | Defect must be contained within a 1 mm circle (include distortion). | |
| | Distortion (this includes line distortion, bull eyes, and press dimples). | None allowed. | |
| | Fixture marks/mold marks. | Allowed up to 10 mm from edge of glass. | |

| Table 14.76 Lamination Inspection Guidelines—Frit | | | |
|---|---|---|---|
| **Title** | **Description** | **Design Rule** | **Consequences & Lessons** |
| Frit | AS1 Mark location | Design ±3.0 mm in all directions.<br>Only 1 AS1 mark is required on each shaded w/s. | Will not meet customer expectations.<br>All rules pertain to "A" zone.<br>Refer to SAE J 198 for zones. |
| | Clear Border | The area between the glass edge and the outer edge of the paint band.<br>4 mm max or per part specification. | |
| | Discolored ceramic paint, light paint | Paint that is reddish brown in color, caused by paint fired too hot or wrong viscosity. None allowed. | |
| | Dot Matrix | The pattern of dots in the silkscreen.<br>Partial Dots:<br>Allowed up to half a dot missing with no more than 4 occurrences.<br>Smeared Dots:<br>Not more than 2 smeared dots, no more than 2 occurrences.<br>Missing dots are not allowed for OEM. | |
| | Heat Grid/Silver Paint Breaks | No broken lines allowed. Breaks less than 1 mm can be reworked. Only 1 reworkable break allowed per line and no more than 2 broken lines allowed per grid.<br>Width tolerance: –0 + .5 mm.<br>Height tolerance: –0 + .5 mm.<br>See part specification. | |
| | Ohms/Amps | Rework: If visually objectionable. | |
| | Paint voids, scrapes, smears, saw tooth | Allowed up to 10 mm from edge of glass. | |

| Table 14.77 Encapsulation—Mold Release | | | |
|---|---|---|---|
| **Title** | **Description** | **Design Rule** | **Consequences & Lessons** |
| Encapsulation / mold release | Encapsulation tool maintenance | Frequent visual tooling inspections | Many suppliers have seen numerous failure modes at many plants and OEMs in the past on encapsulated glass, where the bond is with the encapsulations to a body flange. The failure does not occur on every part, nor is this a problem on every type of encapsulation. The problem usually occurs on approximately every 10th part from a run. It can be mitigated sometimes by visual inspection of the parts and then cleaning with alcohol, and other cleaners. Sometimes, though, the mold release surface pollution is not readily apparent visually. EPDMs are different, and though it can be relatively easy to get a short-term bond between the adhesive and EPDM, longer term adhesion can be problematic. This is because EPDM continues to "weep" to its surface chemical contaminants that compromise the adhesion interface in some cases. |

| Table 14.78 Frit and Primer Dimensions | | | |
|---|---|---|---|
| **Title** | **Description** | **Design Rule** | **Consequences & Lessons** |
| Frit & Primer Dimensions | Clear Primer | Width of primer to $C_L$ of urethane path | 25 mm ± 2 (typical) |
| | Black Primer | Typical width of urethane applies 15 mm | 20 mm ± 2 (typical) |
| | $C_L$ of Urethane bead to Edge of Glass | | 25 mm constant all around typical |

| Table 14.79 Rate of Change in Drop Glass | | | |
|---|---|---|---|
| **Title** | **Description** | **Design Rule** | **Consequences & Lessons** |
| Rate of Change | Framed side lites | 1 mm/100 mm of travel (Green) | Packaging constraints. Tighter tolerances. |
| | Unframed side lites | .5 mm/100 mm of travel (Green)<br>.75 mm/100 mm of travel (Yellow)<br>1 mm/100 mm of travel (Red) | Fewer suppliers able to meet stringent demands. Potential piece price increase due to tighter tolerances. |

| Table 14.80 Glass Deflection upon Installation | | | |
|---|---|---|---|
| **Title** | **Description** | **Design Rule** | **Consequences & Lessons** |
| Glass Deflection Upon Installation | Acceptable glass flex | ≤1 mm to 3 mm (typical) (Green)<br>>3 mm (Red) | Glass breaks in tension |

# Chapter 15
## Design and Processing Criteria

Chapter 14 covered specific design rules. These rules in table form will guide you as the engineer toward a successful launch of your product. Chapter 15 covers global guidelines that address state and federal requirements that must be met to build a saleable product: FMVSS, ECE, and ISO mandated rules. In 15.3 Windshield Design Technical Assumptions, standard verbiage that is used on transmittals between an OEM and a supplier, such as a Statement of Work (SOW), is provided.

## 15.1　Global Comparison: Laminated Windscreens

| Global Comparison Chart (CLEPA): Laminated Windscreens | | | | |
|---|---|---|---|---|
| Test | Europe ECE Regulation 43 | JAPAN Safety Regulations for Road Vehicles, Article 29 | USA FMVSS 205/ANSI Z26.1-1977 Z26.1a-1980 | Draft Global Standard |
| Windscreen Optics | Test on windscreens<br>• Using defined vision areas<br>• At the installation angle<br>• Test method ISO 3538 | Tests on windscreens<br>• Using defined vision areas<br>• At the installation angle<br>• Test method ISO 3538 | Test of 12″ squares which may be cut from the most curved part of the windscreen<br>• No defined vision area<br>• Not tested at the installation angle<br>• Not as ISO 3538 | As ECE R43 |
| Light Transmission | TL ≥ 75%<br>Test procedure ISO 3538 | TL ≥ 70%<br>Test procedure ISO 3538 | TL ≥ 70%<br>Test procedure ISO 3538 | TL ≥ 70%; i.e., as for USA, Japan, and Directive 77/649/EEC. Forward Field of Vision |

*(continued)*

| Global Comparison Chart (CLEPA):  Laminated Windscreens (continued) | | | | |
|---|---|---|---|---|
| Test | Europe ECE Regulation 43 | JAPAN Safety Regulations for Road Vehicles, Article 29 | USA FMVSS 205/ANSI Z26.1-1977 Z26.1a-1980 | Draft Global Standard |
| Light Stability High Temperature Humidity Fire Resistance | Test procedure and evaluation: ISO 3917 250 mm/min | As ECE R43 89 mm/min | Test procedure ISO 3917 Evaluation high temperature and humidity not as ISO 3917 88.8 mm/min | As ECE R43 90 mm/min |
| Impact 227 g Ball | Test procedure ISO 3537 Test at +104°F (+40°C) and –4°F (–20°C) Varying drop heights according to thickness | As ECE R43 | Test procedure ISO 3537 Test at 77°F (25°C) Standard drop height | As ECE R43 Temperatures the same, but two standard drop heights according to temperature |

## 15.2   Global Comparison: Toughened Bodyglass

| Global Comparison Chart (CLEPA):  Toughened Bodyglass | | | | |
|---|---|---|---|---|
| Test | Europe ECE Regulation 43 | JAPAN Safety Regulations for Road Vehicles, Article 29 | USA FMVSS 205/ANSI Z26.1-1977 Z26.1a-1980 | Draft Global Standard |
| Impact Test 227 g Ball | • Test procedure ISO 3537 • Drop heights: thickness ≤ 3.5 mm to 2.0 m thickness > 3.5 mm to 2.5 m • Flat 300 × 300 mm test pieces | • ISO 3537 • Drop heights as ECE R43 • Test pieces as ECE R43 | • Test procedure ISO 3537 • Drop height: 3.05 m • Test pieces as ECE R43 | As ECE R43 Standard drop height: 2.0 m |
| Impact Test 4.99 kg shot bag | No test | No test | No ISO test. Drop height: 2.40 m • Flat 305 × 305 mm test pieces | No test |
| Abrasion Test | No test for the glass surface. If plastic coated, then: Test procedure ISO 3537 | As ECE R43 | • Test procedure ISO 3537 • Carried out on bodyglass requisite for driving visibility | As ECE R43 |

(continued)

| | | | | |
|---|---|---|---|---|
| **Global Comparison Chart (CLEPA):  Toughened Bodyglass (continued)** | | | | |
| **Test** | **Europe ECE Regulation 43** | **JAPAN Safety Regulations for Road Vehicles, Article 29** | **USA FMVSS 205/ANSI Z26.1-1977 Z26.1a-1980** | **Draft Global Standard** |
| Light Transmission | • Test procedure ISO 3538<br>• In areas requisite for driving visibility: $T_L$>70%<br>• In areas not requisite for driving visbility: $T_L$ no lower limit | As ECE R43 | • Test procedure ISO 3538<br>• For passenger cars the $T_L$ limit is > 70%, except for rooflights<br>• For other vehicles the limits are as ECE R43 and Japan. | As ECE R43 |
| Optical Quality | No test | Sidelights requisite for driving visibility | No test | No test |
| Fragmentation | Test procedure ISO 3537<br>• Production parts are broken using a spring loaded centre punch or pointed hammer from 4 defined breaking points.<br>• The minimum particle count allowed is 40 (in any 5 × 5 cm sided square) with an upper limit of 450 for a thickness < 3.5 mm.<br>400 for thickness >3.5 mm<br>• No elongated particles (splines) in excess of 7.5 cm are permitted.<br>• The maximum particle size allowed is 3 cm².<br>• NB: Some deviations on the above are permitted. Example: splines up to 10 cm. | ISO 3537<br>• Requirements are similar to those specified in ECE R43.<br>• Some small differences in the allowed deviations.<br><br>Deviation examples:<br>• Splines up to 15 cm.<br>• In case particle count < 40, then: particle count ≥ 160 in any 10 × 10 cm square is acceptable. | No fragmentation test as ISO 3537. Instead: there is a fracture test on the flat 305 × 305 mm squares used for the 227 g ball test. The drop height is increased until the glass breaks. The weight of the largest fragment must not exceed 4.25 g. No evaluation of the length of fragments.<br><br>*1996 revision of ANSI Z26 proposes a fragmentation test as ISO 3537, with only one defined break position (25 mm inboard of the midpoint of the longest edge).*<br><br>*The interpretation of results is based on the weight of the largest fragment, which must not exceed 4.25 g. This equates to the following maximum particle sizes:*<br>*3 mm thickness: 5.6 cm²*<br>*4 mm thickness: 4.2 cm²*<br>*5 mm thickness: 3.4 cm²*<br><br>*No evaluation of the length of fragments.* | As ECE R43, with some changes:<br>• A single centre break position is specified.<br>• The upper particle count limit is removed. Minimum limit remains at 40.<br>• The elongated particle limit is raised from 7.5 to 10 cm.<br>• Determination of the largest particle weight rather than of the area, e.g. for glass up to 4.5 mm thickness the weight must not exceed 3.0 g. This equates to:<br>3.9 cm² for glass 3 mm.<br>3.0 cm² for glass 4 mm.<br><br>Unlike ECE R43 and Japan, no deviations are permitted. |

## 15.3 Windshield Design Technical Assumptions

| Windshield Design Technical Assumptions | |
|---|---|
| Description | Windshield with shade band |
| Drawing Number | CAD |
| Part Number | TBD3 |
| Construction | 6.76 (3 Solar/0.76 Clear PVB with shade band/3 Clear) ± 0.4 |
| Block Size | 718 × 2454 |
| Edge Finished | Type # 3 |
| Type of Ceramic Paint Band | Solid and dot matrix all around for surface # 4 |
| Upper Width | 25 |
| Lateral Width | 25 |
| Lower Width | 25 |
| Tolerance | ± 1 |
| Frit Clear All Around | Max 5 |
| | Splitting off—when you pull on the urethane, instead of getting cohesive failure of the urethane to the paint, the paint will actually pull off the body causing glass to come out in an accident or roll over. |
| | The paint therefore becomes the weak link in the chain, thus, you don't meet retention requirements for the glass, thus the term, paint splitting off the body. |
| Minimum Dot Size | 1.2 |
| Central SAG & Tolerance | 3.05       –3/+5 |
| Longitudinal SAG | 313.1 |
| Off Form Tolerance All Around | 0 +3 |
| Rate of Change | 2.5/150 |
| Dimensional Tolerance | ± 1 |
| Datum's Side | ± 3 |
| Opposite Datum's Side | |
| Open Primer | Primer application on glass periphery |
| | 435-18: Cleans glass surface |
| | 435-20: Promotes adhesion |
| Construction Transmittance TL% (Light Transmittance) | TL: 73% |

## 15.4 Laminate Windshield Inspection Notes

**"A" Area** is located 75 mm from inner edge of paint band or 100 mm from edge of glass without paint band.

If shade band is greater than 100 mm from edge of glass, the top of the A Area starts at the bottom of the shade band. Not more than one defect within any 300-mm area.

**"B" Area (OEM Windshield)** is all the area located between the A and C Areas.

Not more than two defects within any 300-mm area.

**"B" Area (Replacement Full and Half Windshield)**—On driver side, it is all the area located between the A and C Areas; and on the passenger side, it is the whole area except for the outer 10 mm around the perimeter of the glass.

Not more than two defects within any 300-mm area.

**"C" Area** is located 10 mm from edge of glass.

All defects are allowed unless otherwise specified or as long as the strength of the glass is not impaired.

Defects of less than minimum size are acceptable provided they do not interfere with vision.

Defects not addressed by this specification will be located on glass part specification.

## 15.5 Typical Product Flow from Point of Production to Assembly

Note: When you are looking to take costs off the value chain in glass manufacture and shipping, your options are somewhat constrained. The glass manufacturer's float plant operations will not be moved due to the enormity of scale of the operations. Your processing and assembly offer some hope in the form of regional distribution points. They are small facilities located near the point of assembly (POA).

Ideally, you want to keep your transportation distances as close to the POA as possible.

**Value Stream Mapping**

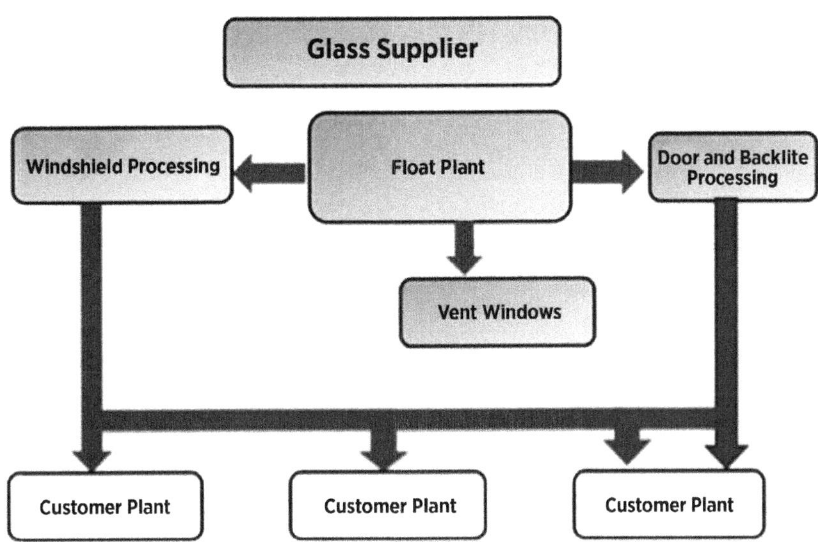

# About the Author

A mechanical engineer and automotive glass specialist, Lyn R. Zbinden began his career with Saturn Corporation in the U.S., where he designed the first Saturn windshield and backlite. He then led General Motors' glass and glazing efforts, with a multinational team responsible for the design, engineering, manufacturing, and sourcing of its product portfolio for the next decade.

Among the many new processes he implemented, one that made history in the industry was the development and successful use of dedicated design software.

The gains in productivity and competitive edge supported by this project made Lyn the General Motors' recipient of the "Outstanding Engineering Achievement of the Year" award in 1997. He was also responsible for the modernization of assembly plants in Europe and Mexico, optimizing plant layout, improving material handling and reducing waste.

His designs can be seen from military vehicles to Formula One race cars.

Lyn R. Zbinden currently consults to the transportation industry, and is an active member of SAE International.